THE BREEDING BIRDS OF EUROPE: A PHOTOGRAPHIC HANDBOOK

The Breeding Birds of Europe 2

A PHOTOGRAPHIC HANDBOOK

Sandgrouse to Crows

Manfred Pforr and Alfred Limbrunner

Translated by Richard Stoneman
Edited by Iain Robertson

CROOM HELM LONDON

Croom Helm Ltd, 2–10 St John's Road, London SW11

British Library Cataloguing in Publication Data

Pforr, Manfred
The breeding birds of Europe.
2: Sandgrouse to crows
1. Birds—Europe—Pictorial works
I. Title II. Limbrunner, Alfred
III. Robertson, Iain
598.294 QL690.A1

ISBN 0-7099-2020-2

The preparation of the text was shared by Alfred Limbrunner and Manfred Pforr.

Colour lithography by Elnain, KG, Wiesbaden
Printed by Heinrich Silber, Niesetal-Heiligenrode
Bound by Freitag and Co., Kassel, West Germany

Contents

Volume II

Sandgrouse (Order Petroclidiformes)

Recent research has shown that the Sandgrouse exhibit a close relationship to Waders (Charadriiformes) rather than to the Pigeons and Doves (Columbiformes). We have chosen to follow Voous (*Ibis*, vol. 115, p. 197) in treating Sandgrouse as a distinct group with the single family Pteroclididae, of which two species breed regularly in Europe.
Sandgrouse are pigeon-like ground-birds with small heads, pointed wings and tails, short feathered legs and a flight similar to that of the Golden Plover. The sexes are dissimilar, with the males being more colourful. Sandgrouse inhabit deserts, semi-desert, steppe and other dry regions. They make long journeys to water, usually in the early mornings. They drink like pigeons, gulping without raising their heads. The water is stored in the crop and the gullet and mouth are used to carry water back to the young. In some species water is carried in the abdominal feathers, which have specially adapted filaments.
Sandgrouse are shy of man and, though gregarious, they nest singly or in very close colonies. The nest is a scrape in the ground and the young are nidifugous and fully downy on hatching. In Europe Sandgrouse are largely resident, though some dispersal occurs.

Pin-tailed Sandgrouse

Black-bellied Sandgrouse

Pallas's Sandgrouse

Pigeons and Doves (Order Columbiformes)

The order Columbiformes consists of one recent family, the Columbidae. The terms 'pigeon' and 'dove' are largely interchangeable, though the former is used mainly for the larger Columba species and the latter for the Streptopelia group. However, both Rock and Stock Doves are typical Columba pigeons. Of about 300 species worldwide only six breed in Europe. All feed on plant material, seeds and grain. This hard food is prepared for digestion by the strong muscular gizzard. Pigeons drink by sucking, a habit shared only with the Sandgrouse and Button Quails. Young pigeons are fed on 'pigeon-milk', a nutritious substance produced by sloughing off the cells of the crop lining. The sexes are basically similar in plumage and the voice is typically a coo-ing noise. European pigeons may be resident, partially migratory or fully migratory, depending upon the species.

Cuckoos (Order Cuculiformes)

The order Cuculiformes contains the Cuckoos (Cuculidae) and the Turacos (Musophagidae). Only the Cockoos occur in Europe, represented by three species, two of which breed. The European species are nest-parasites, laying their eggs in the nests of other species and allowing the foster-parent to rear the young. The females usually specialise in parasitising a particular species, and the eggs, though much larger, are very similar in colour to those of the host species. The European Cuckoos are migratory and are forced to make long-distance migration without the help of their parents. Cuckoos are rather slim birds with long tails, a pointed downcurved bill and two toes forward, two backwards (zygodactile). Most of the 127 species of Cuckoo in the world rear their own young, and they occur in almost all regions of the world except the treeless arctic wastes and the most arid desert areas. The voice of the 'common' Cuckoo is particularly distinctive and is responsible for making it a well-known bird throughout Europe.

Owls (Order Strigiformes)

Although the Owls share a number of anatomical features with Raptors, such as a hooked bill, long curved talons and large, well-developed eyes, they are not closely related.
The order Strigiformes contains two families, the Tytonidae or 'Barn Owls' and the Strigidae, 'genuine owls'. Thirteen species of owl breed in Europe and they are a diverse group, including birds from Starling-size (Pygmy Owl) to the large Eagle Owl, which is nearly two feet tall. The colouring is predominantly grey or brown, usually with spotting or barring. A facial disk is present, most marked in the Barn Owl, and some species have 'ear tufts'. The head is large and extremely mobile; in some species the ears are asymmetric, which allows very accurate pinpointing of sounds.
The dense soft plumage of owls conceals their slim bodies and the softness of the feathers helps to make their flight almost noiseless. A further adaptation is that the outer web of the primaries has a comb-like structure of filaments, preventing the air from whistling against a hard edge in flight.
Owls feed on prey which they catch themselves, varying in size from small insects to large mammals and birds. Stealth is most important in obtaining prey rather than speed or agility.
With one or two exceptions, the European owls are hole-nesters and lay round white eggs. The young have white down at first and hatch asynchronously.
Most owls are able to remain in the same areas throughout the year, though the more northerly species may move to more favourable breeding areas. Only one species, the Scops Owl, is largely migratory.

Nightjar (Order Caprimulgiformes)

This order contains five families; the Oilbirds (Steatornithidae), Potoos (Nyctibiidae), Frogmouths (Podargidae), Owlet-Nightjars (Aegithalidae) and the Nightjars (Caprimulgidae). Only the latter occur in Europe and are represented by two breeding species. Nightjars are medium-sized birds with very large gapes, small feet, narrow wings and long tails. They have a cryptic plumage pattern and soft feathers which enables them to fly silently. Nightjars are active in twilight and at night, feeding on flying insects. They do not build a nest but lay their eggs on bare ground. The plumage of the parents gives sufficient camouflage to protect them from predators. Both species are migratory, there being insufficient insect food to sustain the population in a European winter.

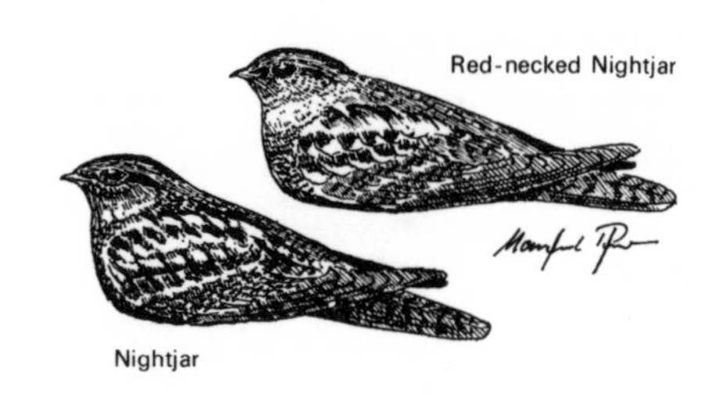

Swifts (Order Apodiformes)

Two families make up this order; Apodidae, the 'true swifts', and Hemiprocnidae, the crested swifts. Only the former is represented in Europe, where four species breed. The swifts are similar in appearance to swallows and martins. They have long sickle-shaped wings, slim bodies and short tails. They have a fast, sweeping flight and catch insects on the wing.

Swifts have dark brown or blackish plumage with some white on the underside or on the rump. The sexes are alike in plumage. The feet are small and poorly developed; all four toes point forwards and are adapted to clinging to vertical surfaces.

Swifts are gregarious birds; they nest in colonies and feed in flocks. The nests are built of débris that the birds collect in flight and are cemented together with saliva.

Swifts are entirely migratory, moving south in winter to areas where insect-life is more abundant.

Kingfishers and their allies
(Order Coraciiformes)

The order Coraciiformes is made up of nine families, four of which are represented in Europe. These are the Kingfishers (Alcedinae), Bee-eaters (Meropidae), Rollers (Coraciidae) and Hoopoes (Upupidae), each with one species breeding in Europe.
The order contains a highly diverse group of birds whose brightly coloured plumage adds a tropical touch to the European avifauna. All are hole-nesting species which lay white eggs. The Bee-eater is a sociable bird, breeding in colonies and feeding in flocks. The rest are rather solitary birds. With the exception of the Kingfisher they are migratory.

Woodpeckers (Order Piciformes)

This is a large order of birds found throughout the world with the exception of Australasia and Madagascar. Six families occur within the order; Jacamars (Galbulidae), Puffbirds (Bucconidae), Barbets (Capitonidae), Hoenyguides (Indicatoridae) and Woodpeckers (Picidae). Only the latter is found in Europe.
Of 210 species worldwide, ten breed in Europe including the Wryneck, which is part of the distinct sub-family (Jynginae).
The true Woodpeckers are highly adapted for life in trees. The claws are strong and well developed for climbing and the toes are arranged with two pointing forwards and two backwards (in the Three-toed Woodpecker the vestigial first toe has been lost). The tail feathers are stiff and act as a support when climbing tree-trunks. The bill is strong and pointed for chiselling out insects from bark and dead wood and for excavating the nest-hole. The tongue is extremely long with a barbed tip and it is covered with a glue-like secretion. This enables the birds to extricate larvae from holes deep in the wood.
The species range in size from that of a Sparrow to that of a Crow. The plumage is basically similar in both sexes, though there is usually a difference in the pattern of red on the head.
The flight is markedly undulating, with the wings held against the body between flaps. The species are all hole-nesters, excavating their own nest chambers, and the eggs are pure white.
The Wryneck differs from all the 'true' woodpeckers in having a cryptic plumage which is totally alike in both sexes. The tail is not stiffened and pointed and the species does not excavate its own nest but uses a natural cavity. It is the only wholly migratory bird in the group.

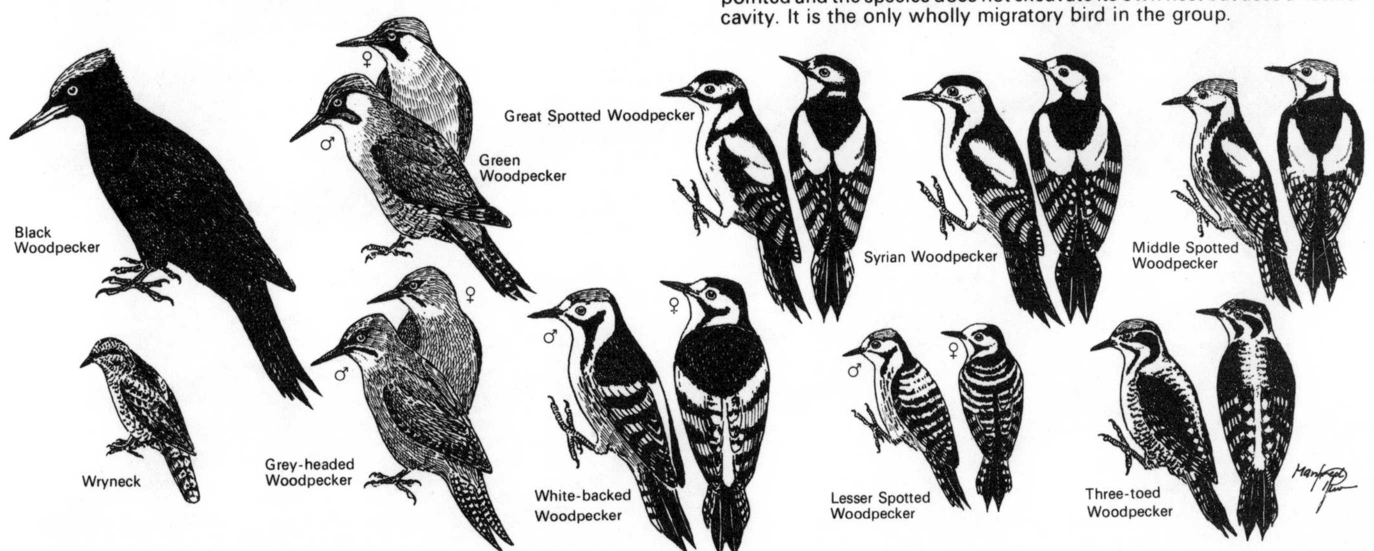

Perching birds and Songbirds

(Order Passeriformes)

This is the richest order in the world in terms of species. There are over 4,000 species, more than half the birds in the world. In Europe about 165 breed regularly, ranging in size from the Goldcrest, which is the smallest European bird, to the Raven.
The most important feature in the order is the structure of the larynx, which facilitates a particularly wide range of sounds from the vocal apparatus.
The musical ability of many of the songbirds is highly developed and, though both sexes have the same anatomical features for singing, it is generally only the males which sing. In many species there is considerable sexual dimorphism.
The nests are complex structures, often very beautifully constructed of delicate materials. The young are hatched naked or sparsely covered in down and remain helpless in the nest for a relatively long period.
The various families are distinguished by the shape and structure of the bill, which reflects their feeding habits, as well as the great differences in song and calls. The various families are found in every habitat except the open sea. Many are adapted to a terrestrial life, obtaining their food and building their nests on the ground. Others live almost exclusively in trees. Members of the order Passeriformes make up the largest part of the European avifauna and are the most numerous and most commonly met with birds. A large proportion of the species migrate to take advantage of the availability of insect food in the tropics.

Larks Alaudidae

Swallows and Martins Hirundinidae

Swallow

Red-rumped Swallow

House Martin

Sand Martin

Crag Martin

Pipits and Wagtails Motacillidae

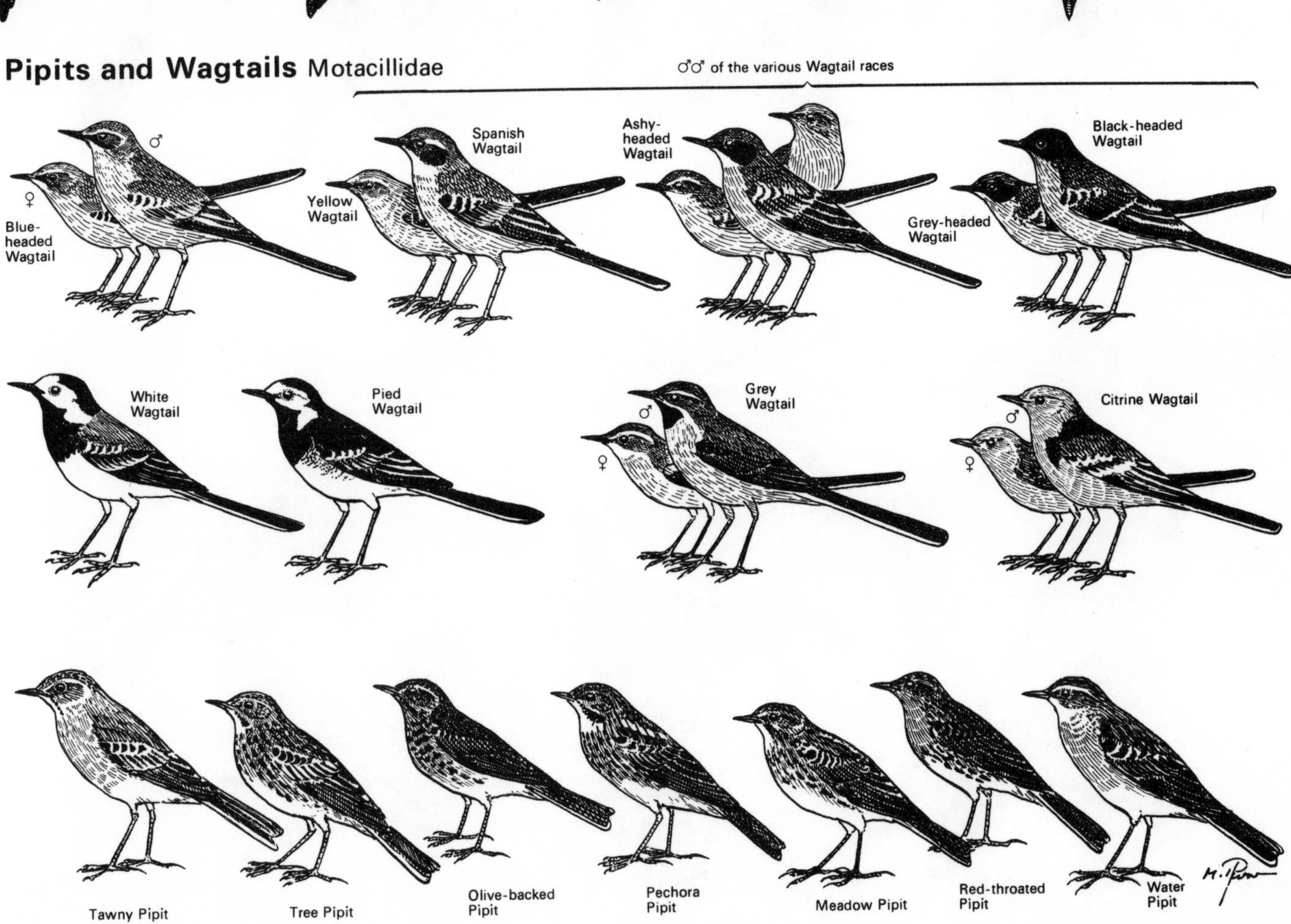

Shrikes Laniidae

Waxwings
Bombycillidae

Wrens
Troglodytidae

Dippers
Cinclidae

Accentors
Prunellidae

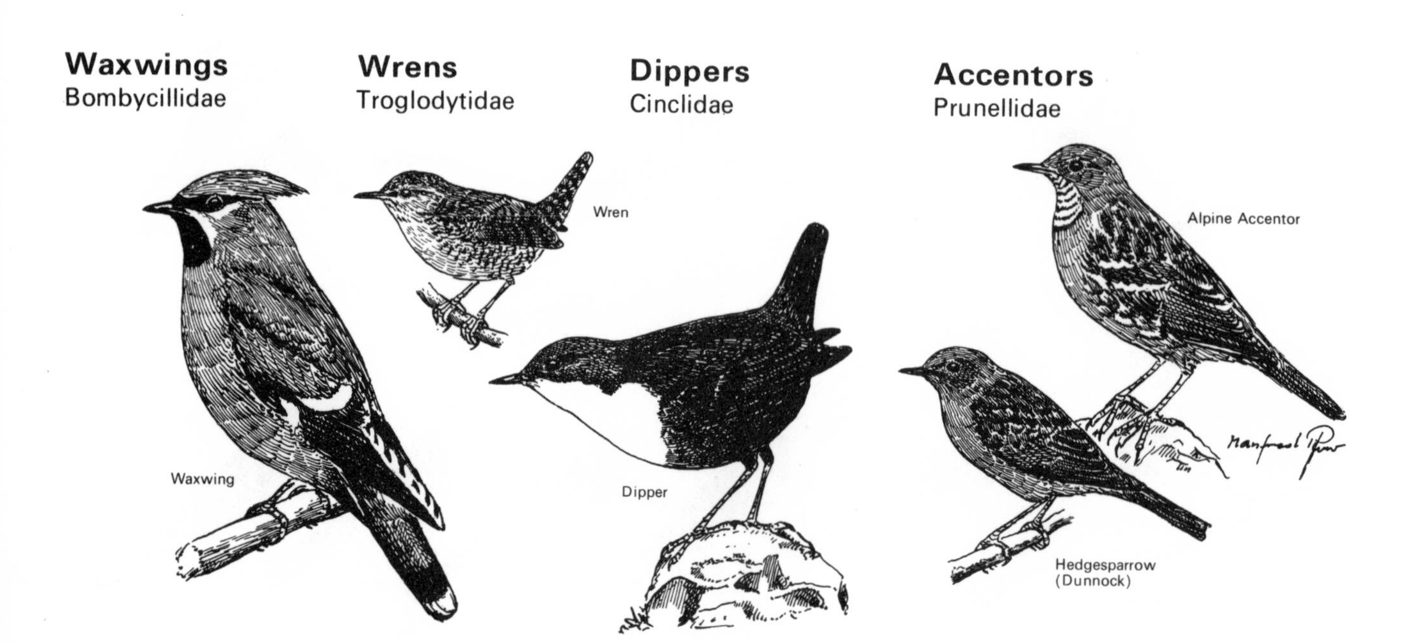

Warblers Sylviidae

Warblers Sylviidae

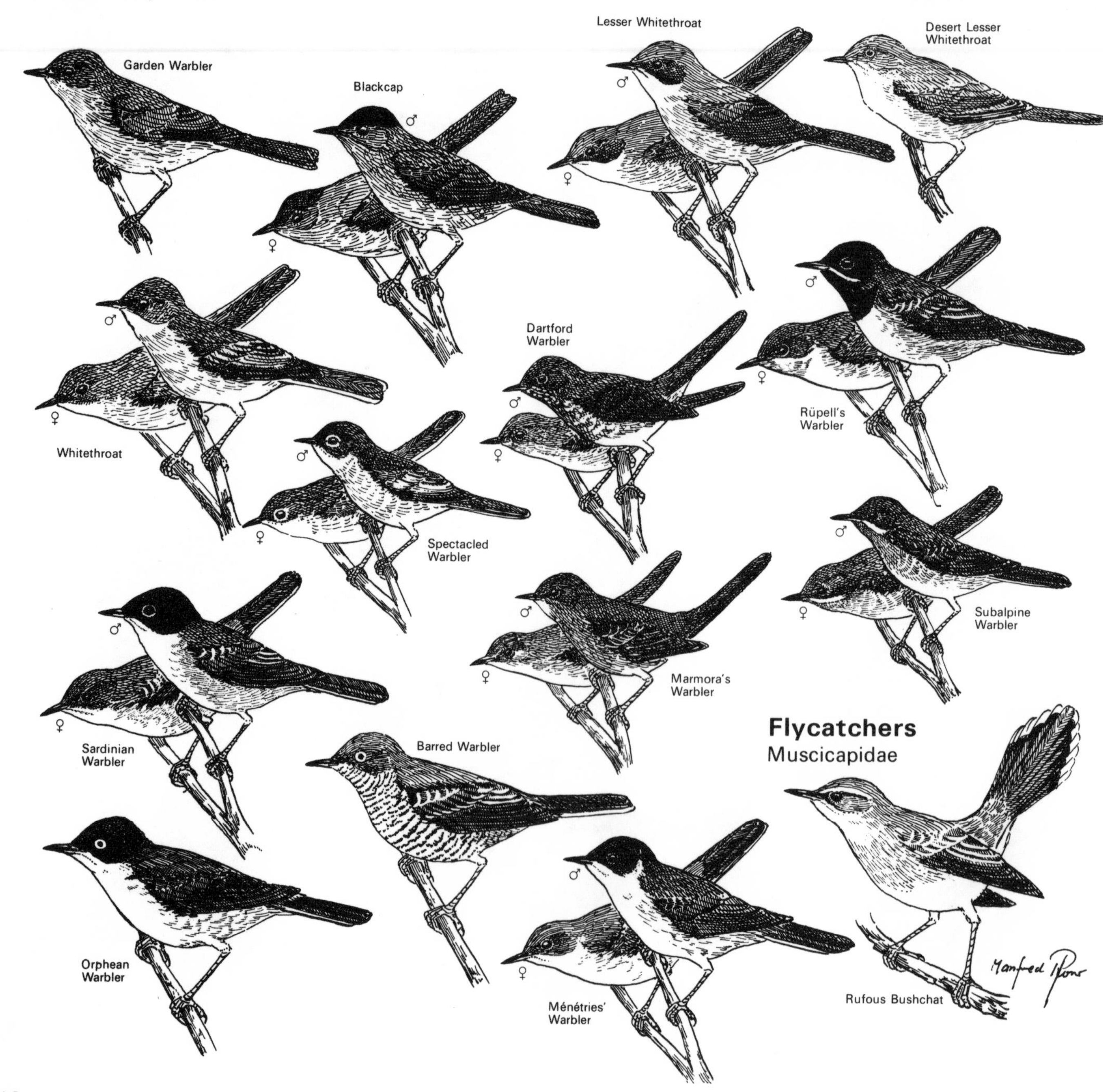

Warblers Sylviidae

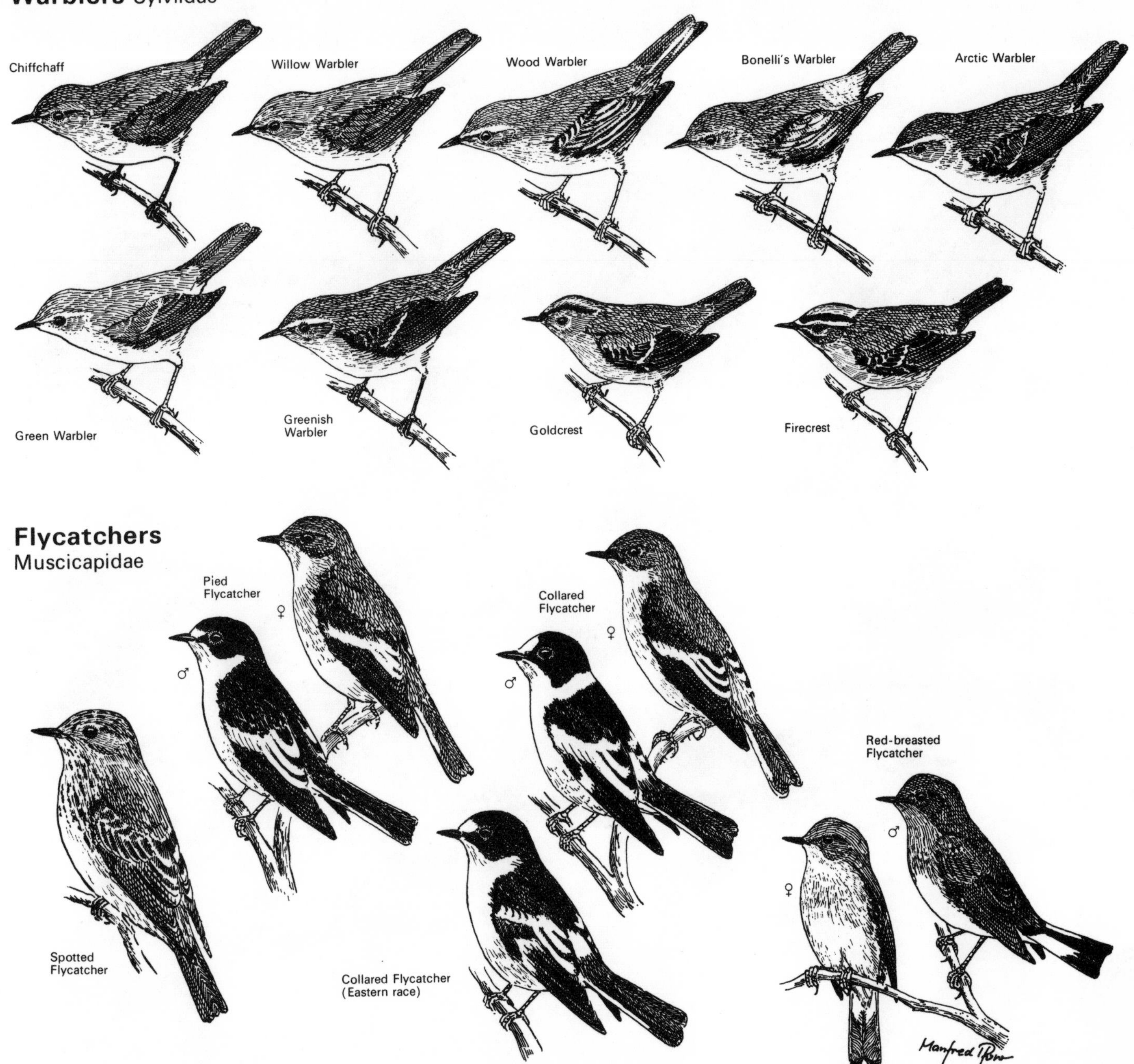

Flycatchers Muscicapidae

Red-flanked Bluetail
♀
♂
Thrush Nightingale (Sprosser)
Nightingale
Red-spotted Bluethroat
♂
White-spotted Bluethroat
♀
♂
Robin
♀
Black Redstart
♂
♀
♂
♀
Siberian Rubythroat
♂
Redstart
Whinchat
♀
♂
♀
Stonechat
♂
♂
♀
Rock Thrush
♂
Blue Rock Thrush
Manfred Pforr

Flycatchers Muscicapidae

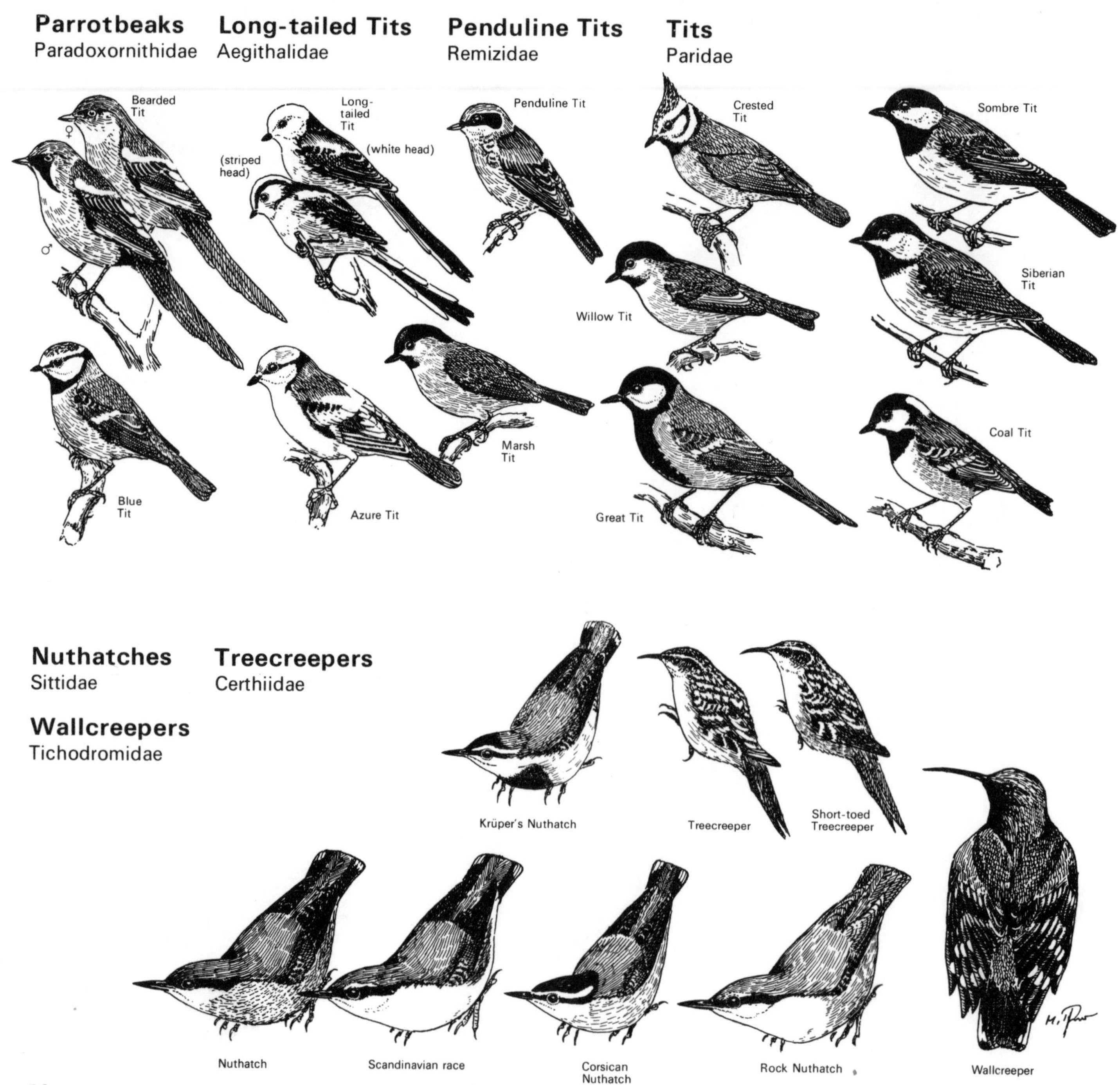
Parrotbeaks
Paradoxornithidae
Long-tailed Tits
Aegithalidae
Penduline Tits
Remizidae
Tits
Paridae
Bearded Tit
♀
♂
Long-tailed Tit
(white head)
(striped head)
Penduline Tit
Crested Tit
Sombre Tit
Willow Tit
Siberian Tit
Blue Tit
Azure Tit
Marsh Tit
Great Tit
Coal Tit
Nuthatches
Sittidae
Treecreepers
Certhiidae
Wallcreepers
Tichodromidae
Krüper's Nuthatch
Treecreeper
Short-toed Treecreeper
Nuthatch
Scandinavian race
Corsican Nuthatch
Rock Nuthatch
Wallcreeper

Buntings Emberizidae

Finches Fringillidae

Sparrows Passeridae

Orioles Oriolidae
Starlings Sturnidae

Siberian Jay
Jay
Nutcracker
Azure-winged Magpie
Magpie
Alpine Chough
Carrion Crow
Hooded Crow
Chough
Rook
Jackdaw
Raven

Special Section

The Breeding Birds of Europe

Pin-tailed Sandgrouse

(Pterocles alchata)

Pin-tailed Sandgrouse is the commonest of the two European species. It inhabits dry plains, semi-desert, steppe and areas of arid vegetation. The diet is mainly the leaves and seeds of plants, including cereal crops in some areas. The birds fly to water in the early morning, large numbers gathering together when conditions are dry. The ♂♂ carry water in the down of their breast-feathers back to the young chicks. Though markedly gregarious when flying to water and when feeding, the birds usually nest solitarily, though loose colonies occur. The nest is a shallow scrape of natural depression, usually without lining. It is often near a tuft of vegetation. The incubation and care of the young is shared by both parents. The adult birds sit very tight on the nest, allowing their natural camouflage to protect them. Increased cultivation and reclamation of arid areas has reduced the European population of this species.
Pin-tailed Sandgrouse are resident throughout their European range, though some local dispersal takes place after the nesting season.

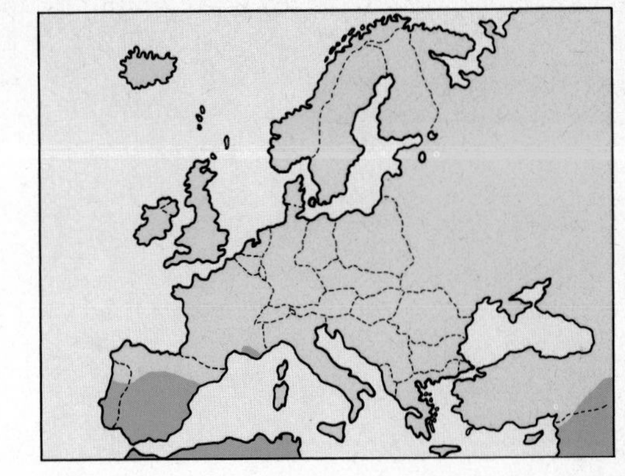

Length:	32 cm
Wing length:	19.9–21.5 cm
Weight:	225–290 g
Voice:	In flight: 'ketarrh-ketarrh', 'arrr-a-arr'
Breeding period:	End of April, May. 1 or possibly 2 broods per year. Replacement clutch possible
Size of clutch:	3 (2) eggs
Colour of eggs:	Buffish with dark flecks
Size of eggs:	45.7×29.8 mm
Incubation:	19–23 days, beginning from 2nd egg
Fledging period:	Nidifugous: independent at 10–14 days, able to fly at 1 month

Ad. ♀ in breeding plumage, reservation photograph (OvF)

3-day-old young, La Crau, Southern France (OvF)

Turkey, June 1977 (Wa)

left) Brooding ♀, crouching, Turkey, June 1977 (Wa)

Black-bellied Sandgrouse

(Pterocles orientalis)

The Black-bellied Sandgrouse is found in the same dry habitat as the Pin-tailed Sandgrouse, with which it often mixes. It feeds entirely on vegetable matter such as seeds and leaves. During the breeding season birds associate as pairs, flocks only occurring at regular drinking places. In winter quite large flocks occur where food is abundant.
The ♂♂ will fly long distances to obtain water for the young. They wade into the water to soak their breast-feathers, then fly back to the nest, where the young drink the water from the feathers.
No nest is made, merely a shallow scrape by the bird. It relies on its cryptic plumage for camouflage and will crouch motionless rather than fly off when approached. Both sexes share in incubation when rearing the young.
Though mainly resident in the western part of the range, the eastern populations are partially migratory, moving south in winter. There is some evidence to suggest that change in habitats due to increased agricultural development has reduced the population in western Europe.

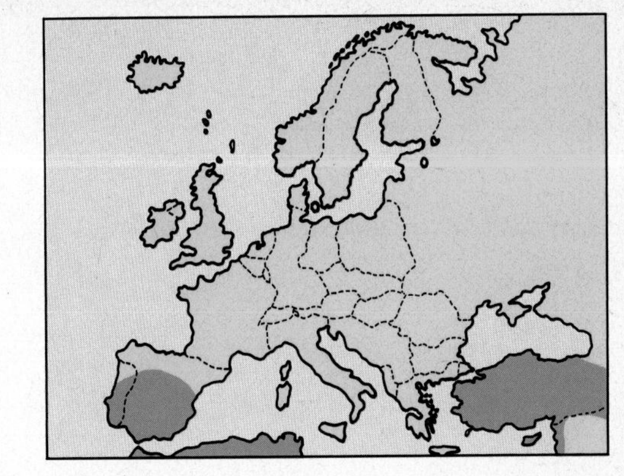

Length:	35 cm
Wing length:	♂ 23.7 cm, ♀ 22.6 cm
Weight:	♂ 428 g, ♀ 383 g
Voice:	In flight: 'chyrr-yrr-yrr', repeated several times
Breeding period:	Mid-April, May, June. 1–2 broods per year. Replacement clutch probable
Size of clutch:	3 (2) eggs
Colour of eggs:	Yellowish to light grey-brown, with darker brown irregular patches
Size of eggs:	47.9×32.5 mm
Incubation:	21–22 days, beginning from 1st egg
Fledging period:	Nidifugous; feeding themselves after leaving the nest

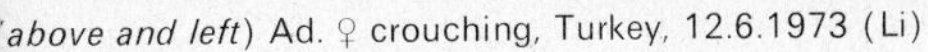

(above and left) Ad. ♀ crouching, Turkey, 12.6.1973 (Li)

Chick crouching, Turkey, 12.6.1973 (Li)

Turkey, 31.5.1974 (Li)

Stock Dove (Columba oenas)

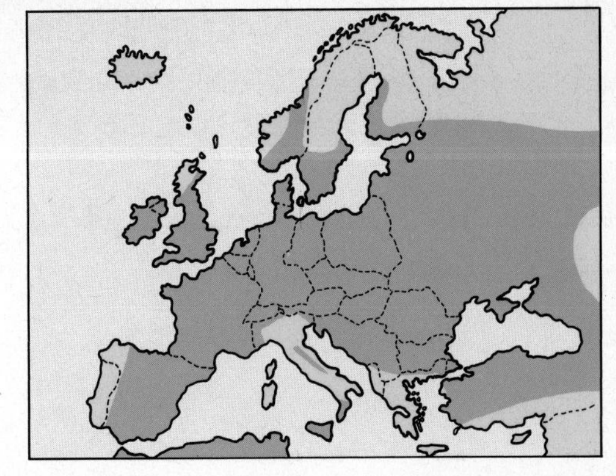

Stock Doves are found in areas of open woodland, broad-leaved forest, parks, gardens and open country with scattered trees. They also occur on coastal and inland cliffs, sand-dunes, old buildings and ruins.
The diet is mainly vegetable matter, mostly leaves of *brassicas*, seeds and grain; some animal matter such as small snails may also be taken.
Not an abundant bird, it is usually seen in small parties, though flocks of several hundred may occur where food is plentiful. Stock Doves have a circling display-flight, often gliding with raised wings. Wing-clapping takes place but is soft and feeble compared with that of the Woodpigeon.
The nest-site is usually in a hollow tree or hole in a tree, old building or cliff. Sometimes rabbit burrows are used, and the species takes readily to nest-boxes. Nests are often unlined but sometimes a layer of twigs or dead leaves is used. Both sexes share with incubation and care of the young. The ♀♀ may begin incubation of the second clutch leaving the ♂♂ to care for the young of the preceding brood.
Stock Doves are largely resident in western Europe though birds from north and east of the range migrate to winter in south-west Europe.
Migration: Mainly September–November, with return passage in February–April.

Length:	33 cm
Wing length:	21.0–22.7 cm
Weight:	270 g
Voice:	Courting, 'ooo-*woo*'
Breeding period:	April–August. 2 or sometimes 3 broods
Size of clutch:	2 eggs
Colour of eggs:	White
Size of eggs:	37.8 × 29 mm
Incubation:	16–18 days, beginning from 1st egg
Fledging period:	Nidicolous; leaving nest at 27–28 days

eft) Ad. ♀, reservation photograph (Li)

Juv. in nesting hole, reservation photograph (Li)

5-day-old nestling, reservation photograph (Li)

d. ♂, reservation photograph (Li)

Rock Dove (Columba livia)

The Rock Dove is the ancestral form of the domestic or feral pigeon. It is now most numerous as the familiar 'street-pigeon' of towns and cities everywhere. The natural habitat is rocky sea-cliffs, mountainous areas and inland cliffs. Food consists of a variety of seeds and grain as well as snails and littoral débris. The Rock Dove is generally seen singly or in small parties, though larger flocks may occur where the species is numerous.
The nest-site is in caves and rock crevices, the feral form breeding on ledges on buildings, ruins and man-made structures. The nest varies from a few stems and strands of grass to a substantial structure of roots, bents, brass and even seaweed. Nests may be single or in small colonies. Both sexes share in incubation and care of the young.
In the wild form Rock Doves are now confined to remote coastal areas of north-west Europe and the wilder, mountainous regions of southern Europe. In many places the population has been diluted with feral birds and a wide range of plumage types exist. Rock Doves are largely resident and sedentary, though some dispersal of northern birds takes place in winter.

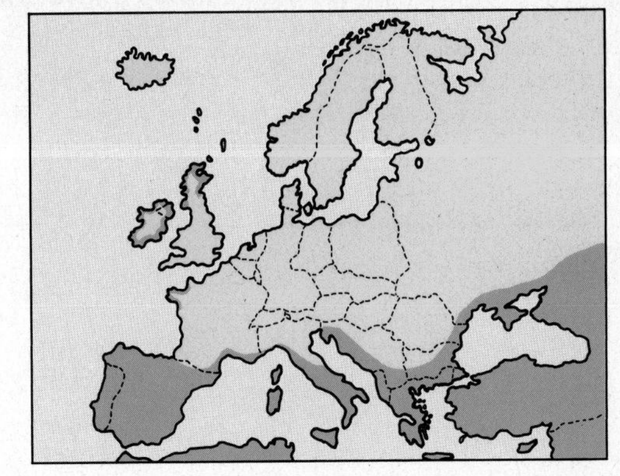

Length:	33 cm
Wing length:	21.0 × 23.2 cm
Weight:	260 g
Voice:	Coos like domestic pigeon: 'ooo-ooo-ooo, oor-oor-roo-woo'
Breeding period:	Western populations breed all year, eastern ones from end-March to end-July, 2–3 times. Replacement clutch frequent
Size of clutch:	2 (1) eggs
Colour of eggs:	White
Size of eggs:	39 × 28.8 mm
Incubation:	17–19 days, beginning from 1st egg
Fledging period:	Nidicolous; remaining 4–5 weeks in nest and then able to fly

bove and left) Ad., Turkey, June 1978 (GI)

Turkey, June 1978 (GI)

Croatia, Yugoslavia, June 1980 (Pf)

Woodpigeon (Columba palumbus)

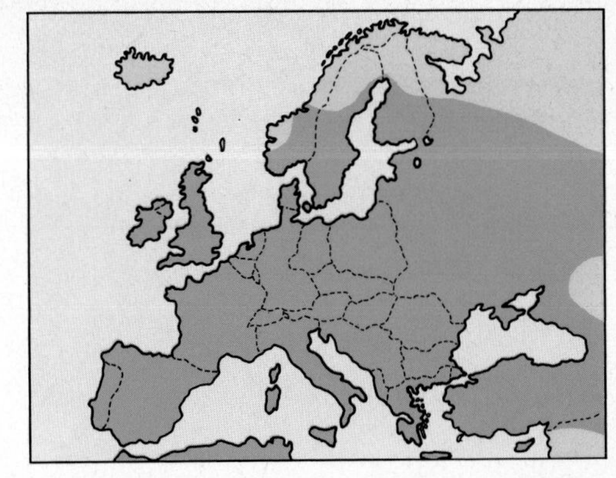

This is the largest and most numerous of the European pigeons. Once confined to forest and woodland it is now common in open country with clumps of trees, in parks and gardens. It has become tolerant of human presence and is now at home in towns and cities.
Woodpigeons feed mainly on vegetable matter, especially leaves of *brassicas* and grain. When present in large numbers the species can do considerable damage to crops and constitutes a major pest, controlled by shooting.
Though solitary when nesting, the Woodpigeon is gregarious on migration and in winter. Flocks of many thousands occur where food is plentiful. The Woodpigeon has an undulating display flight, often clapping the wings loudly at the top of a glide. The nest is a slight but firm structure of twigs, with the eggs often visible from below. The usual site is in a tree or bush, though ledges on buildings may be used and sometimes the old nest of another species is utilised. Both sexes share in incubation and care of the young.
The northern and eastern populations are migratory, wintering in south-west and south-east Europe. British birds are almost entirely sedentary.
Migration: Autumn passage in September–December with 'hard-weather movements' in winter. Spring return in February to early June.

Length:	40 cm
Wing length:	23–25 cm
Weight:	500 g
Voice:	Courting call: 'croo-croo croo-croo coo'
Breeding period:	April to July. 2–3 broods per year. Replacement clutch possible
Size of clutch:	2 eggs
Colour of eggs:	White
Size of eggs:	44.2×29.5 mm
Incubation:	15–17 days, beginning from 1st egg
Fledging period:	Nidicolous; leaving the nest at 4 weeks, but only able to fly after a further week

d. with nest material, Bavaria, June 1978 (Pf)

Wintering individual, Bavaria, January 1960 (Li)

Bavaria, May 1978 (Pf)

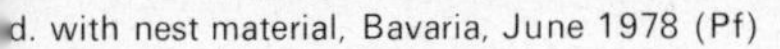

eft) Feeding, Bavaria, May 1965 (Li)

Turtle Dove (Streptopelia turtur)

The Turtle Dove prefers low broad-leaved woodland, parkland, copses, hedgerows and large gardens. It is less attached to dense woodland than the Woodpigeon.
The diet is almost entirely seeds and grain, though some leaves, shoots and animal-matter are taken.
Turtle Doves are often seen perched on wires or picking up grit at the roadside, usually in pairs or small groups. On migration and in late summer large flocks occur.
The species has a display-flight, rising steeply then gliding downwards. 'Injury-feigning' has been recorded from nesting birds.
The nest is usually a thin platform of twigs, grass and stems and is usually sited lower in the vegetation than those of other pigeons. Sometimes the old nest of another species, or a squirrel's drey, is used. Both sexes share in incubation and rearing the young. Turtle Doves are entirely migratory, wintering in a belt across tropical Africa south of the Sahara. Some individuals may overwinter in Europe, often in company with Collared Doves in the vicinity of a grain store which guarantees adequate food.
Migration: Departs breeding areas in August–September, returning in late March to May.

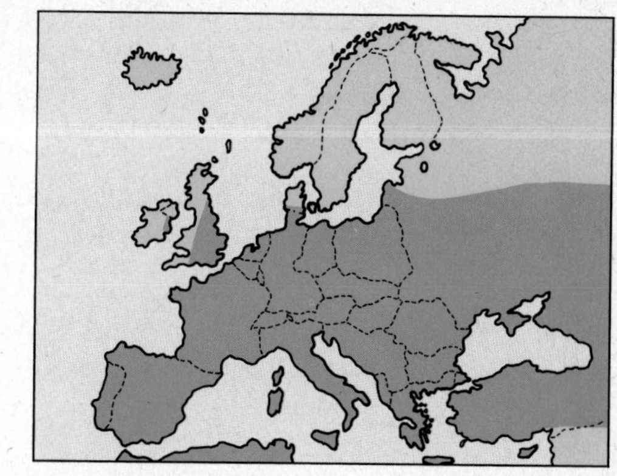

Length:	27 cm
Wing length:	17.3–18.2 cm
Weight:	160 g
Voice:	Courting call: a throaty 'trurr-trurr-trurr'
Breeding period:	Mid-May, June, beginning of July. 1–2 broods per year. Replacement clutch possible
Size of clutch:	2 eggs
Colour of eggs:	White
Size of eggs:	30.7×23 mm
Incubation:	13–15 days, beginning from 1st egg
Fledging period:	Nidicolous; able to fly at *ca* 20 days, independent at 30–35 days

I. ♂ brooding, Austria, 1969 (Li)

Nestlings, Turkey, June 1973 (Li)

Turkey, May 1973 (Li)

ft) Imm., Bavaria, 1977 (Li)

Collared Dove (Streptopelia decaocto)

Originally an Asian species, the Collared Dove spread through Asia Minor into south-east Europe, colonising western Europe this century. The species bred in Britain for the first time in 1955. It is now widespread and a common breeding bird throughout much of western Europe. It is closely associated with human habitation, and is found in parks, gardens and farmland.
It feeds mainly on seeds and grain, taking advantage of grain stores, chickenfeed, etc. to such an extent that it has become a pest in some areas. The distinctive mating call is uttered from birds perched on street-lamps, television aerials and rooftops. The nest is a sparse platform of twigs and leaves sited in a tree, often an evergreen. Both sexes share in incubation and in rearing the young. The nesting period is often extended and this results in the ♂♂ caring for the young of one brood whilst the ♀♀ begin incubation of another. Interbreeding with the Turtle Dove has been recorded.
Collared Doves are largely resident and sedentary. Movement does occur in spring and early summer, and this may indicate that there is a continuing expansion of the range. Large numbers occur together in winter.

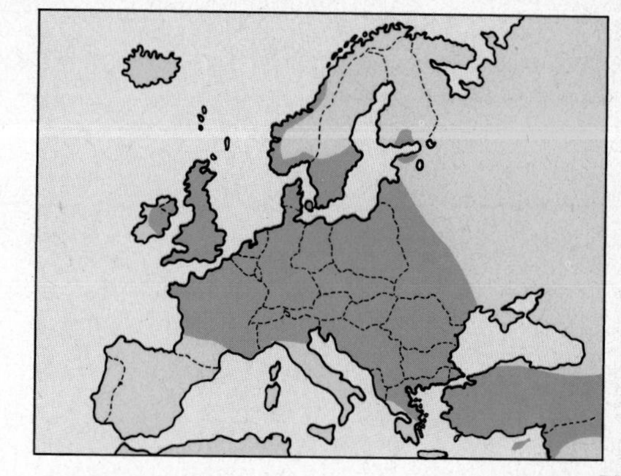

Length:	28 cm
Wing length:	17–18.3 cm
Weight:	200 g
Voice:	Male mating call: 'coo-oooo-cul', also 'kuurr'
Breeding period:	March to September, rarely in winter. 3 or more broods per year
Size of clutch:	2 eggs
Colour of eggs:	White
Size of eggs:	30×23.2 mm
Incubation:	14–16 days, beginning from 1st egg
Fledging period:	Nidicolous; fledged at 18–21 days, but are still tended by the ♂ while the ♀ starts a second clutch.

varia, November 1976 (Pf)

Juv., Lower Saxony, West Germany, July 1961 (Sy)

Lower Saxony, June 1961 (Sy)

ft) Ad., Bavaria, February 1976 (Pf)

Cuckoo (Cuculus canorus)

The Cuckoo is found in almost every type of habitat, from open parkland to forest, reed-beds, moorland and scrub tundra. The diet is exclusively insects, particularly caterpillars and other larvae. It will take the large, hairy caterpillars which are rejected by most birds. Cuckoos are beneficial in taking many harmful insect pests. The diet of the young Cuckoo varies according to the foster-parent. Cuckoos do not form a pair-bond; several ♂♂ will display to one ♀, and the ♀♀ are polyandrous. The ♀ lays one egg in the nest of the host species. Before laying, she removes one of the host's eggs, which she may eat. Each ♀ specialises in one host species, perhaps the species by which it was raised. Over sixty different host species are known in Europe, Dunnock, Robin, Meadow Pipit, Reed Warbler and Pied Wagtail being some of the commoner ones.

The young Cockoo squirms beneath the eggs or young of the host species and ejects them from the nest. This ensures that there is no competition for food.

Cuckoos are entirely migratory, wintering throughout most of Africa south of the Sahara.

Migration: Adults leave European breeding grounds in July and August, the young following mainly in August and September. Spring passage takes place in late March to late May.

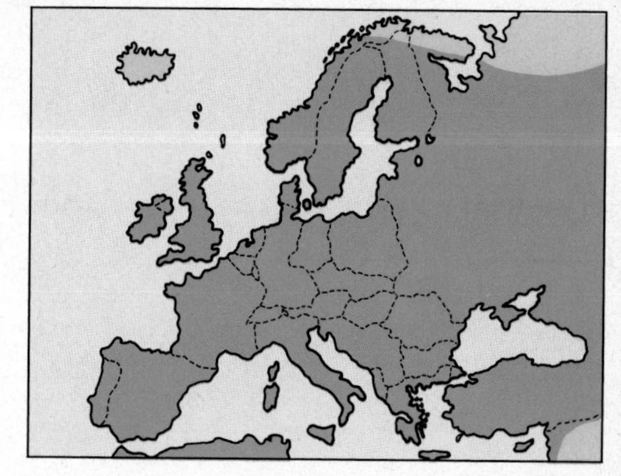

Length:	33 cm
Wing length:	21.6–23 cm
Weight:	100 g
Voice:	Courting call of ♂: 'cuck-koo' and 'kakakoo'; also a harsh puffing. ♀: a shrill bubbling note
Breeding period:	Dependent on the host bird: Mid-April, May, June, July
Size of clutch:	8–12 eggs, 1 egg in each host bird's nest
Colour of eggs:	Uniquely variable, because it approximates, often excellently, the colouring of the host bird's eggs
Size of eggs:	23×17.2 mm
Incubation:	12.5 days
Fledging period:	Nidicolous; leaves nest at *ca* 21 days but is fed by host parents for a further three weeks

v. with host bird (Spotted Flycatcher), Bavaria, 1968 (Li)

Juv., 4 days old, June 1978 (Pf)

Cuckoo's egg in Reed Warbler's nest, Bavaria, June 1977 (Pf)

ft) Fledged young Cuckoo, Bavaria, 1978 (Pf)

Great Spotted Cuckoo

(Clamator glandarius)

The Great Spotted Cuckoo is found in open woodland, olive groves and riverine bushes. Though the species is probably of African origin it breeds in Iberia, southern France and south-east Europe. It feeds mainly on large insects and larvae, including the hairy caterpillars.
Like the (Common) Cuckoo, this species is a nest-parasite, specialising in corvids, particularly the Magpie, Azure-winged Magpie and Carrion/Hooded Crow. No pair-bond is formed and after mating the ♀ lays her eggs in the nests of the host species, usually removing one of the host's eggs before laying. Sometimes several ♀♀ will lay eggs in the same nest and the young Cuckoos are reared along with the remaining young of the foster parent. At other times the host's young are ejected or smothered by the young Cuckoos.
Great Spotted Cuckoos are migratory; a few overwinter in extreme southern Europe but most appear to winter in Africa north of the Equator. The wintering range in Africa is not precisely known due to the presence of resident Great Spotted Cuckoos breeding in that continent.
Migration: Adults depart the European breeding areas as early as June, with young following in July and August. Spring migration takes place in March and April.

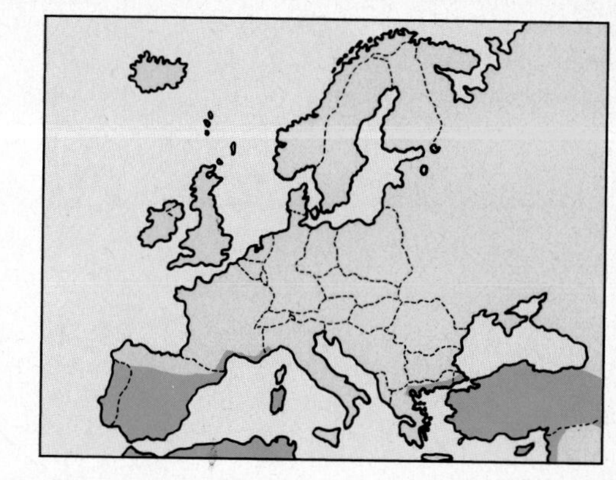

Length:	39 cm
Wing length:	19–22 cm
Weight:	200 g
Voice:	'Keeow-keeow-keeow-keeow', rasping
Breeding period:	End of April, May, June
Size of clutch:	In one season 10–20 eggs
Colour of eggs:	Light greenish-blue with numerous light brown and grey specks
Size of eggs:	32×24 mm
Incubation:	12–14 days, shorter than host species
Fledging period:	Nidicolous; young birds can climb out of the nest at 15 days, and are fledged at 20–24 days

♂, La Crau, Southern France, May 1967 (Li)

Greece, Nestos Delta, June 1977 (Sy)

Great Spotted Cuckoo's eggs, 1 magpie's egg (dark), Turkey, 14.6.1977 (Li)

ft) Nestling in Hooded Crow's nest, Greece, 26.5.1978 (Li)

Barn Owl (Tyto alba)

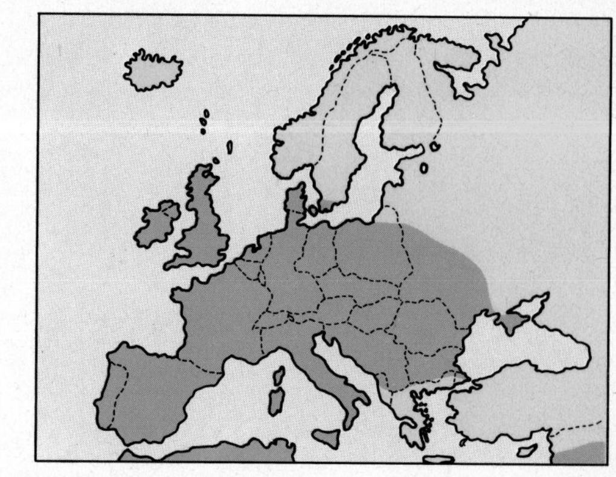

One of the most cosmopolitan of birds, the Barn Owl is found in farmland and similar open country. It is most active by night, but may be seen at dusk or even in broad daylight as it hunts over fields. It feeds mainly on small mammals such as mice, rats and shrews. Small birds, beetles, reptiles and amphibians are also taken at times. It has a regular hunting beat, flying low over the ground to drop silently onto prey.

The Barn Owl nests in hollow trees and buildings of all types, particularly farm outbuildings, church-towers and the like. It will also nest in caves and rock crevices and will take to nest-boxes. Incubation is by the ♀ alone, the ♂ providing food for her and the young. No nest material is used though a scrape may be made in existing débris and nests may have a litter of pellets in and around them. The young have two coats of down, unlike other owls. The first white coat is replaced by a buffish one at about 12 days.

There are two forms of Barn Owl in Europe, the white-breasted race (*T. a. alba*), which is commonest in Britain, Ireland and parts of northern Europe, and the dark-breasted form (*T. a. guttata*), which is more frequent in eastern and southern Europe.

The Barn Owl is mainly resident and sedentary, though young birds may disperse considerable distances.

Length:	34 cm
Wing length:	26–30 cm
Weight:	300 g
Voice:	Usual note is long, eerie screech; also variety of hissing and snoring noises
Breeding period:	Mid-April, May, June, July, sometimes as early as February, sometimes as late as December, often double-brooded
Size of clutch:	4–7 eggs, laid at 2-day intervals
Colour of eggs:	White
Size of eggs:	39.7×31.6 mm
Incubation:	30–34 days, beginning from 1st egg
Fledging period:	Nidicolous; fledged at *ca* 2 months; independent at about 3 months

. with prey, Lower Saxony, West Germany, July 1971 (Pf)

Immature individual, Bavaria (Li)

Reservation photograph (Li)

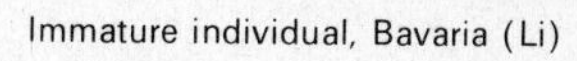

t) Pair in nesting box (*T. a. guttata*), Reservation photograph (Li)

Scops Owl (Otus scops)

The Scops Owl is found in open woodland, parkland, hedges and large gardens in southern Europe. It feeds mainly on insects, especially beetles, large moths and grasshoppers. Small birds are rarely taken. It feeds almost exclusively at night.

During the day Scops Owls are difficult to see; they remain well concealed in thick cover, often well camouflaged against a tree-trunk. At night the chief indication of their presence is the persistent, monotonous call which is uttered from a perch in a tree.

The nest-site is usually in a hole in a tree, often an abandoned woodpecker's hole, or in a hole in a wall or building. Sometimes the old nest of another bird may be utilised. No material is used to line the nest. Incubation is by the ♀ alone, the ♂ providing food which is fed by the ♀ to the young.

Scops Owls are entirely migratory, wintering in a broad belt across Africa south of the Sahara. Though solitary or occurring in pairs in Europe, the species has been seen in small groups in the winter quarters.

Migration: Departs breeding grounds in September–October, returning in April–May.

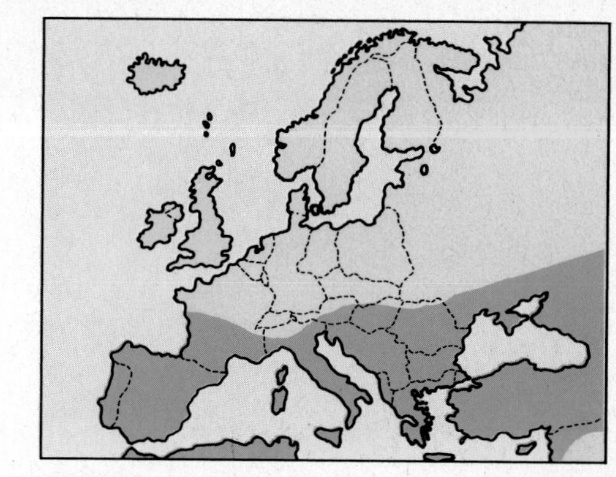

Length:	19 cm
Wing length:	14.5–16 cm
Weight:	100 g
Voice:	Courting call of ♂: 'Kiu', repeated often; ♀ replies in a higher, more disyllabic note; the two may often be heard in turn
Breeding period:	End of April, May, mid-June. 1 brood per year
Size of clutch:	4–5 eggs (3–6)
Colour of eggs:	White
Size of eggs:	31.3×27 mm
Incubation:	24–25 days, begin from complete clutch
Fledging period:	Nidicolous; young climb out of hole at three weeks, fly at 4 weeks, independent at 7 weeks

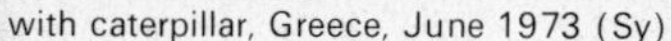

with caterpillar, Greece, June 1973 (Sy)

Juv. in nest, Greece, June 1973 (Sy)

Reservation, Ludwigsburg, West Germany 1975 (Schw)

*) Ad. at daytime roost, Turkey, 18.5.1975 (Li)

Eagle Owl (Bubo bubo)

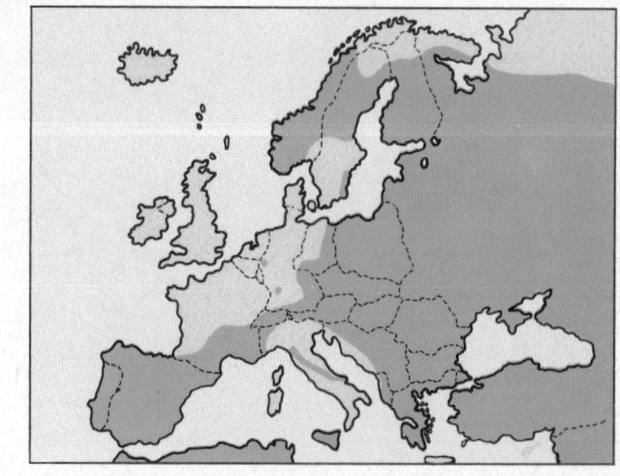

One of the largest European owls, the Eagle Owl occurs in a variety of habitats including extensive forest, mountains, river valleys, rocky cliffs and even large reed-beds. It hunts at dusk and by night, taking an assortment of prey; mammals from Roe Deer downwards to mice, birds up to the size of Capercaillie and reptiles, fish and large insects. It will not tolerate other birds of prey in its territory and as a result the diet often includes birds such as Long-eared and Tawny Owls, Kestrel and Buzzard.
The species has a loud, deep mating call which can carry several kilometres and is a good indication of its presence.
The nest-site is usually in a rock-crevice or cave, a hollow tree, sometimes on the ground or in the nest of another large bird. No nesting materials are used. Incubation is by the ♀ alone and she broods the young for the first three weeks or so, while the ♂ provides the food. When the young are larger both parents hunt. The Eagle Owl is resident and largely sedentary, though it sometimes occurs outside the breeding range in winter. In parts of the European range the species is afforded special protection, with nest-sites being guarded to prevent disturbance or theft of young or eggs.

Length:	66–71 cm
Wing length:	♂ 42–47 cm, ♀ 45–50 cm
Weight:	2,000 g
Voice:	Courtship call: a deep 'oo-hu'; also 'kveck kveck'. Young beg by snarling hoarsely
Breeding period:	Mid-March, April, in southern Europe even in February. 1 brood per year. Replacement clutch possible
Size of clutch:	2–3 (4–6) eggs
Colour of eggs:	White
Size of eggs:	59.3×48.3 mm
Incubation:	34–36 days, beginning from 2nd egg
Fledging period:	Nidicolous; young may leave nesting place at 5 weeks, not yet able to fly; at 14 weeks they can fly well

d. ♀ mantling over young, Bavaria (Li)

Juv. in nest, Bavaria (Li)

Bavaria (Li)

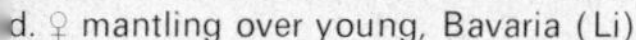

eft) Pair with young at eyrie, reservation photograph (Li)

Snowy Owl (Nyctea scandiaca)

The Snowy Owl is an inhabitant of the arctic tundra and fells of northern Europe. It feeds on mammals up to the size of Arctic Hare, though lemmings are one of the most important items. It will also take birds up to Ptarmigan size, particularly unfledged young. The dependence upon lemmings in much of the range means that the breeding cycle of the owls varies according to the abundance of food. In a good lemming year large clutch-sizes are known, while in a bad year birds may not nest at all.
It hunts by flying low over the ground, or waiting on an exposed perch to drop onto prey. Most hunting takes place at dusk or at night, though there is no darkness in the Arctic summer.
The nest-site is on the ground, often on a hummock or other raised site giving a clear view around the nest. The nest is a shallow scrape, sometimes lined with moss and a few feathers. Incubation is by the ♀ alone, the ♂ providing food which she feeds to the young. When the young are well grown both parents hunt. The ♂ often stands guard near the nest and is aggressive in defence of the territory; it may attack domestic animals and man.
Snowy Owls may disperse from the breeding areas in winter, depending to a large extent on availability of food. Sometimes large irruptions take place, with birds occurring well south of the breeding range.

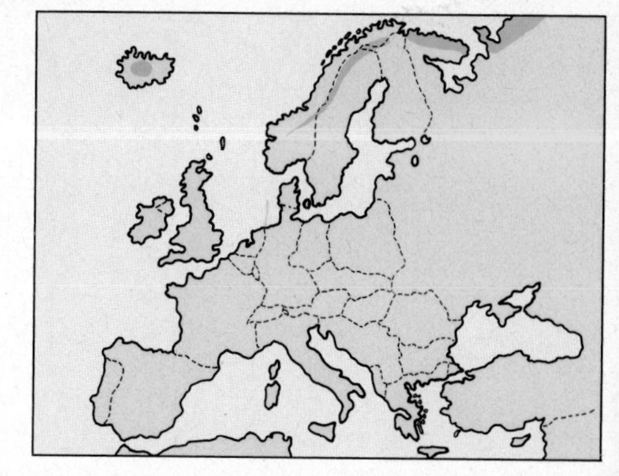

Length:	54–66 cm
Wing length:	♂ 39–41 cm, ♀ 42–46 cm
Weight:	♂ 1,500 g, ♀ 2,000 g
Voice:	♂: 'Vooh' at courting, also barking and quacking sounds. ♀: 'Kekeke' or 'kivkivkiv'
Breeding period:	Mid-May, June. 1 brood per year
Size of clutch:	4–10 (11–15) eggs
Colour of eggs:	White
Size of eggs:	56.3×44.8 mm
Incubation:	32–33 days, beginning from 1st egg
Fledging period:	Nidicolous; at about 14 days the young wander away from the nest; able to fly at 33–50 days

d. ♀, Finland (Hau)

Young birds, 25 days old (Scher)

Chick, 2 days old (Scher)

ft) Ad. ♀, Voliere, Austria (Tö)

Hawk Owl (Surnia ulula)

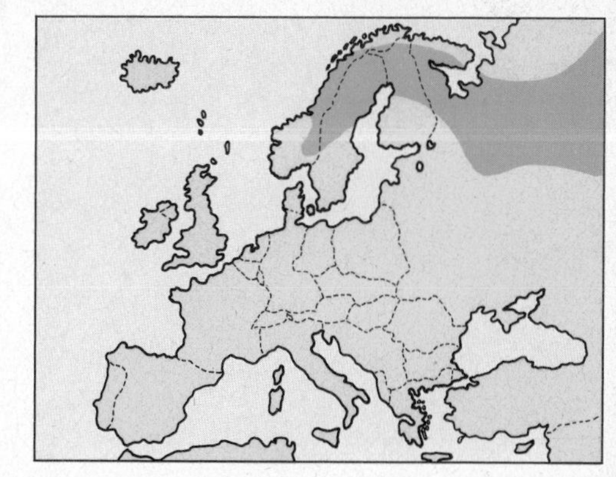

The Hawk Owl inhabits coniferous (preferring larch to dense spruce) and mixed birch forests in northern Europe. The species is active by day, feeding mainly on small mammals, particularly lemmings. Some birds up to the size of Willow Grouse are also taken. The Hawk Owl hunts by dropping onto its prey from a vantage point such as a pole or tree-top. It will also hover in a Kestrel-like manner. They are not shy of man and may attempt to drive off intruders at the nest.

The nest-site is in a hole in a tree, often the abandoned nest of a Black Woodpecker. The old nests of corvids and other birds may be used. There is a preference for hollows in broken branches or tree-stumps. No nest-lining is used.

Incubation is mainly by the ♀ though the ♂ may take some part. Both sexes hunt for food for the young.

Hawk Owls are mainly resident and sedentary, though some winter-dispersal takes place with birds occurring well to the south of their breeding areas. This is usually the result of a food scarcity, especially after an abundance of small mammals has resulted in a high population density.

Length:	36–40 cm
Wing length:	22.5–24.5 cm
Weight:	300 g
Voice:	At breeding-place a barking 'vik-vik' and shrieking 'quikquikquik'
Breeding period:	Beginning of April, May. 1 brood per year
Size of clutch:	5–6 (3–7) eggs, depending on food
Colour of eggs:	White
Size of eggs:	40×32 mm
Incubation:	25–30 days, beginning from 1st egg
Fledging period:	Nidicolous; leaving the nest at *ca* 25 days, but tended rather longer

d. ♂ at nesting-hole, Finland (Hau)

Imm., Finland (Hau)

Juv., Finland (Hau)

'eft) Ad. ♀ at nesting-hole, Finland (Hau)

Pygmy Owl (Glaucidium passerinum)

The Pygmy Owl is a secretive bird, inhabiting the light coniferous and mixed forests of northern Europe. It dislikes the dense spruce forests, preferring more open woodland. In the south of its range it has a montane distribution. The smallest of the European owls, only the size of a Skylark, the Pygmy Owl feeds on small passerine birds, small mammals and insects. It hunts by both day and night, though it is normally well concealed in the tree-tops or in hollow branches by day. If discovered by small birds it is mercilessly mobbed by them. The nest-site is normally a hole in a tree – conifer, birch or poplar. It favours the abandoned holes made by woodpeckers and will take to using nest-boxes in some areas.
No nesting material is used. The ♀ broods alone; incubation does not commence till the clutch is almost complete, which assures little variation in the development of the young. The ♂ provides all the food for both the ♀ and young.
The Pygmy Owl is resident and sedentary throughout its European range. Some altitudinal movements may occur in winter.

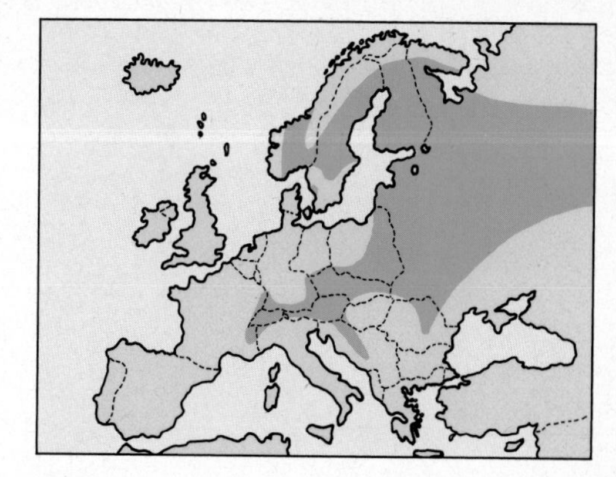

Length:	16 cm
Wing length:	9.8–10.7 cm
Weight:	♂ 60, ♀ 75 g
Voice:	Monotonous whistle: 'kuvih' and 'tyoo-tyoo-tyoo'. Also an ascending series of 'tyoo's, like a scale
Breeding period:	End of April, May. 1 brood per year. Replacement clutch possible
Size of clutch:	5–6 (4–8) eggs
Colour of eggs:	White
Size of eggs:	28.5×23.4 mm
Incubation:	28 days, beg. before clutch complete
Fledging period:	Nidicolous; leaving the nest at 28–30 days

♀ with prey, Sweden (Sw)

Juv., Finland (Hau)

Reservation, Ludwigsburg, West Germany (Schw)

) Ad. ♂, Bavaria (Zeim)

Little Owl (Athene noctua)

The Little Owl occurs in a wide variety of lowland habitats ranging from farmland and open country to riverine bushes, parks, cliffs and quarries. Because it is active in the daytime it is one of the most familiar owls of the region.

It feeds on small mammals, birds, insects, reptiles and even earthworms. It is a fierce hunter, often taking species of birds larger than itself. Sometimes it will kill more prey than it requires. The Little Owl is often seen perched on fence-posts, wires, rocks and buildings. When disturbed it makes characteristic bobbing movements before taking flight.

The nest-site is usually a hole in a tree, a crevice or hole in a cliff or quarry, or in a building. Sometimes it uses a burrow in the ground. No nest materials are used. The ♀ incubates alone, though both parents care for the young.

A number of pale forms occur in south-east Europe, North Africa and the Middle East.

Little Owls are resident and sedentary. The species was introduced to Britain from the late 1880s and it has become common and widespread since then.

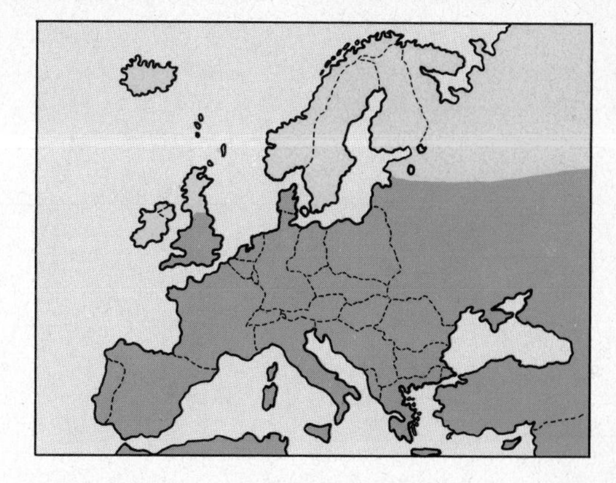

Length:	22 cm
Wing length:	15.2–16.7 cm
Weight:	170 g
Voice:	'Kiew, kiew' and 'gooohk'; otherwise 'kveeauh' and 'koo-vit'
Breeding period:	End of April, May. 1 brood per year. Replacement clutch possible
Size of clutch:	4–5 (3–7) eggs
Colour of eggs:	White
Size of eggs:	35.1 × 29.4 mm
Incubation:	26–29 days, usually beg. with 1st egg
Fledging period:	Nidicolous; young leave nest at *ca* 26 days, but are only able to fly well after a further week

♂ on watch, Austria, 1966 (Li)

Young birds, Greece, June 1976 (Pf)

♀ and clutch, Lower Saxony, West Germany, May 1979 (Sy)

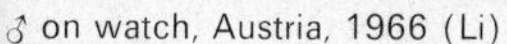

) Ad. (light race), Turkey, June 1974 (Pf)

Tawny Owl (Strix aluco)

The Tawny Owl is found in broad-leaved and mixed woodland, parkland with old hollow trees, riverine areas and parks and gardens in towns. It takes a wide variety of prey; small mammals including voles, shrews, mice and rats, birds up to the size of Pheasant, though mainly passerines, as well as fish, amphibians and insects, It hunts at night, mostly in the cover of trees, taking prey from the ground. During the day it remains concealed against a tree-trunk. If its presence is discovered by small birds it is persistently mobbed. It is bold and aggressive in defence of its nest and is known to strike intruders. The nest-site is usually in a hollow tree though nest-boxes are readily accepted. It will occasionally nest in a rock-crevice and may use the old nest of a corvid or other large bird. It is known to evict Little Owl from a nest-site. No nest-lining is used. Incubation is by the ♀ only with the ♂ providing food, though both parents will hunt when the young are about 3 weeks old. Two colour-phases exist; a brown form and a grey type. the latter is frequent in continental Europe but very scarce in Britain. Tawny Owls are generally resident and sedentary.

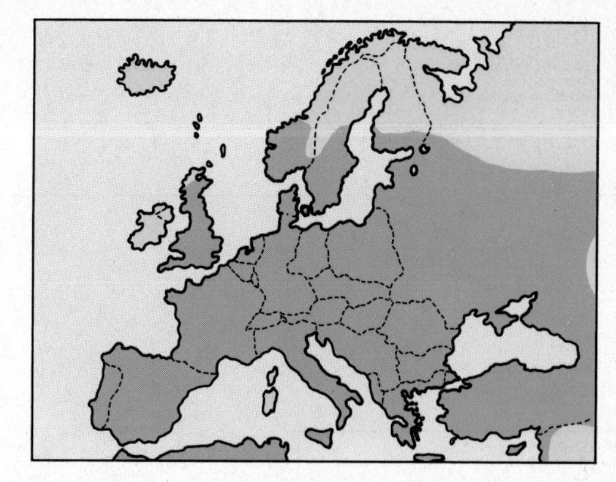

Length:	38 cm
Wing length:	♂ 24.5–26.5 cm, ♀ 25–27.5 cm
Weight:	♂ 450 g, ♀ 600 g
Voice:	Courting call of ♂: 'whoo-hoo-hooh', vibrating at end; otherwise a shrieking 'tu-wit'
Breeding period:	Mid-February, March, April. 1 brood per year. Replacement clutch possible
Size of clutch:	3–5 (2–6) eggs
Colour of eggs:	White
Size of eggs:	48.2×38.4 mm
Incubation:	28–30 days, beginning from 1st egg
Fledging period:	Nidicolous; leaving the nest at 4–5 weeks, but dependent for a further 3 months

grey phase, Bavarian Forest, December 1974 (Pf)

Juv., brown phase (*left*), grey phase (*right*), May 1975 (Pf)

Clutch in nest-box, Bavaria, 10.4.1975 (Li)

Ad., brown phase, Bavaria, 19.4.1977 (Li)

Ural Owl (Strix uralensis)

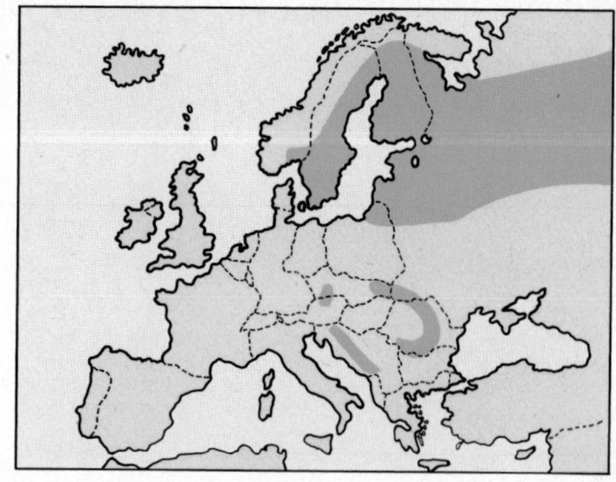

The Ural Owl is found in dense mixed forest and upland coniferous woodland. In Europe it generally requires large areas of undisturbed forest (though some birds will live in city parks) and this is a frequent habitat in Siberia.

The Ural Owl is active by both day and night, hunting in forest-clearings and along rides. It feeds on a variety of small mammals and birds up to the size of Crow. The availability of small mammals has a marked effect on the breeding cycle, the clutch size varying according to abundance of food. It may not breed at all in years when food is scarce.

The nest-site is in a hollow tree or in the old nest of a large bird of prey. Occasionally a cliff-ledge or crevice is used and ground-nests are known. It will take to using nest-boxes at times. No nest materials are used and incubation is by the ♀ alone, with the ♂ providing food. The species is aggressive in defence of the nest-site and territory.

Ural Owls are mainly resident throughout the European range but a southward dispersal may occur in hard winters when food is scarce.

Length:	60 cm
Wing length:	35–38.5 cm
Weight:	♂ 800 g, ♀ 900 g
Voice:	♂ 'whoo-hoo, hoo-hoo', a barking 'khau-khau' and a 'ke-wick' note harsher than that of Tawny Owl
Breeding period:	Mid-March, April. 1 brood per year. Replacement clutch possible
Size of clutch:	2–4 (5–6) eggs
Colour of eggs:	White
Size of eggs:	45×39 mm
Incubation:	27–29 days, beginning from 1st egg
Fledging period:	Nidicolous; leaving nest at 4 weeks, able to fly at 5 weeks, but led for a longer time

t nest with young (Hau)

Fledgeling, Finland (Hau)

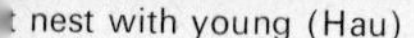

r) Ad., nature reserve, Bavarian forest, January 1974 (Pf)

Great Grey Owl (Strix nebulosa)

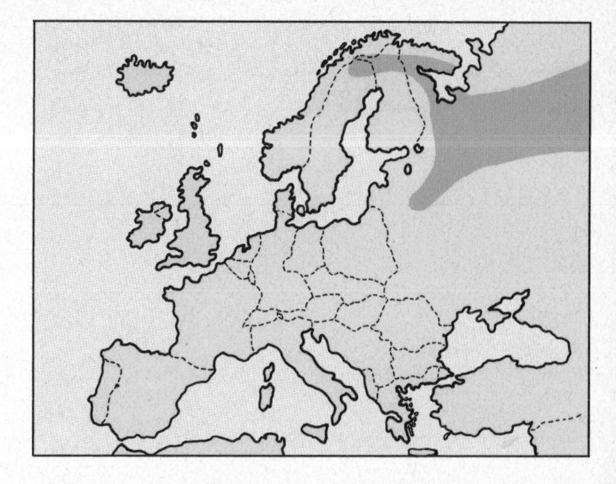

The Great Grey Owl lives in the taiga forests of northern Europe. It is a secretive bird and is thinly distributed through its breeding range.
Despite being one of the largest European owls it feeds on small mammals, mainly voles and shrews. It will also take birds up to the size of Willow Grouse, some amphibians and insects. The species often hunts by day, flying low over the ground in forest-clearings and dropping onto its prey. The dependence upon small mammals means that clutch-size varies considerably with the cyclical abundance of the prey.
The usual nest-site is in the hollow stump of a broken-off tree, though sometimes the old nest of a large bird of prey is used. No nest-lining is provided. Incubation is by the ♀ alone with the ♂ provisioning both the ♀ and the young.
The Great Grey Owl is largely resident and sedentary throughout the European range.

Length:	70 cm
Wing length:	♂ 43.5 cm, ♀ 45.5 cm
Weight:	♂ 800–1,000 g, ♀ 1,100–1,300 g
Voice:	♂ deep 'Oowa-oowa-hoo-hoo-hoo', ♀ 'Neeow'. Also scratching and wheezing sounds
Breeding period:	Mid-April, May. 1 brood per year
Size of clutch:	4–5 (2–3, 6–9) eggs depending on food
Colour of eggs:	White
Size of eggs:	54×43 mm
Incubation:	Unknown; broods from 1st egg
Fledging period:	Nidicolous; leave nest at 3–4 weeks, do not fly well till 5 weeks, may remain under parental care for some months

brooding, Finnish Lapland, 8.6.1977 (Gr)

Ad. ♀ at nest, Finland (Hau)

Zoo photograph (Li)

') Ad. with eggs and young, Finnish Lapland, 8.6.1977 (Gr)

The Long-eared Owl (Asio otus)

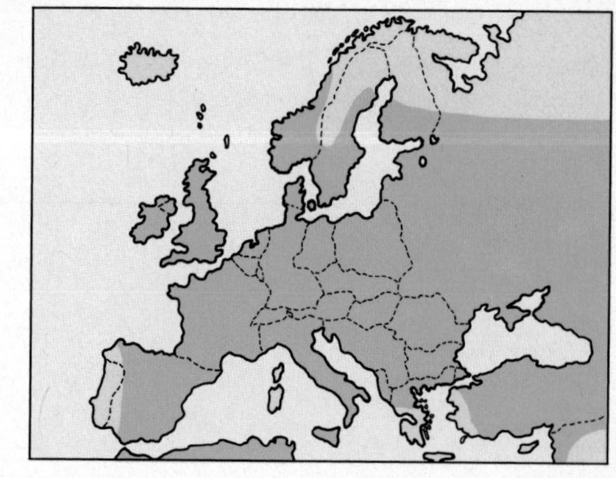

The Long-eared Owl occurs in both deciduous and coniferous woodland, copses, riverine trees, parks and large gardens. It is occasionally found in areas of open heath or moorland.

The diet is mainly voles, mice, rats and other small mammals. Birds are also an important item, particularly finches and Jay. Some insects, fish and amphibians are also taken. The Long-eared Owl is active mainly by night, usually spending the daytime concealed against the trunk of a tree. When discovered it will stretch itself upright, appearing very thin in an attempt to look inconspicuous.

On migration and in winter the species will often hunt in daylight, and at such time it may occur in small groups. During the breeding season it is mainly solitary.

The nest-site is usually in the abandoned nest of a crow, Magpie or other large bird. It will also nest on the ground in some areas. Incubation is by the ♀ alone, the ♂ providing food for both ♀ and young.

The more northerly populations are migratory, wintering in Britain, Ireland and central Europe southwards. The extent of the movements depends upon the availability of food. In some years large irruptions occur and birds may gather at winter-roosts.

Migration: Autumn movements mainly September–November, with return passages in March–May.

Length:	35 cm
Wing length:	27.5–30.5 cm
Weight:	300 g
Voice:	Usual call and song: 'ooo-ooo-oo'. Food-call of young 'feee'
Breeding period:	February onwards. 1 or, rarely, 2 broods per year. Replacement clutch possible
Size of clutch:	4–6 (3–8) eggs
Colour of eggs:	White
Size of eggs:	40.2×32.3 mm
Incubation:	25–30 days, beginning from 1st egg
Fledging period:	Nidicolous; leaving nest at 23–26 days; independent at 2–3 months

ng birds, Bavaria, May 1971 (Pf)

Chicks with unfertilised egg, Bavaria, 25.4.1971 (Li)

Bavaria, April 1969 (Pf)

) ♀ at nest, Bavaria, 1.5.1971 (Li)

Short-eared Owl (Asio flammeus)

Short-eared Owls are found in open areas of marsh, heath, moorland, sand-dunes and rough pasture. In winter it is often seen on arable land. The diet is mostly small mammals, particularly voles. Small birds and some insects are also taken.

The species feeds mainly at dusk but it is frequently seen in daylight. It hunts by quartering low over the ground with a slow, wavering flight. On migration and in winter it may occur in small groups. During 'vole-plagues' quite large numbers may be seen in suitable habitat.

The nest is a shallow, unlined hollow on the ground, usually in the shelter of reeds, heather or other vegetation. Incubation is by the ♀ alone. The ♂ provides food which the ♀ then feeds to the young. The clutch-size is variable, depending upon the availability of small mammals. A second brood may be reared when food is abundant.

The Short-eared Owl is dispersive and migratory, at least in the north of its range. Large-scale movements take place when the population-level is high and food scarce. Many northern and eastern birds winter from Britain and Ireland to the Mediterranean; others migrate as far as the tropics.

Migration: Autumn movements mainly August–October, with return in March–May.

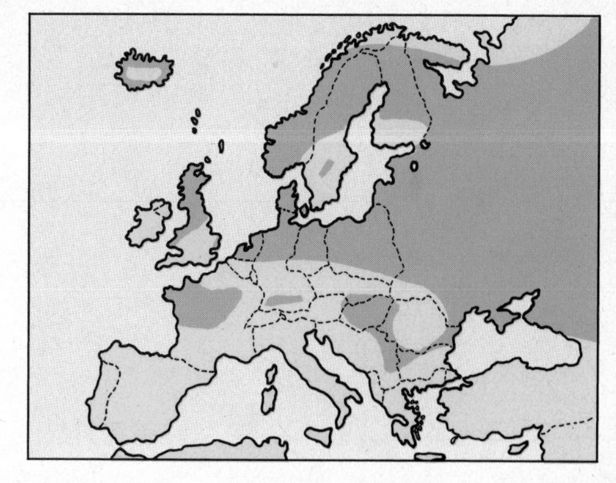

Length:	38 cm
Wing length:	29–33 cm
Weight:	400 g
Voice:	Courting call of ♂: a hollow 'bubububu'; also barking and mewing sounds
Breeding period:	End of March, April, May, June. 1–2 broods per year. Replacement clutch possible
Size of clutch:	4–6 (7–10) eggs
Colour of eggs:	White
Size of eggs:	39.5×31 mm
Incubation:	25–28 days, beginning from 1st egg
Fledging period:	Nidicolous; leaving the nest at 12–17 days, unable to fly for further 10 days

., Finland (Hau)

Nestlings at different stages, Austria, 1968 (Li)

Austria, Hansag, 1968 (Li)

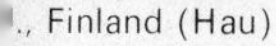

t) ♀ at nest, Austria, 1968 (Li)

Tengmalm's Owl (Aegolius funereus)

The Tengmalm's Owl is found in the dense coniferous forests of the north, mainly in dark spruce taiga. It will also occur in mixed birch and pine forest and in the south of its range it has a largely montane distribution.
The species feeds on small mammals, mostly mice and shrews, as well as on small birds and insects.
During the day it remains well concealed in thick cover, often in the crown of a conifer. It is extremely difficult to see in daylight. It hunts by night along rides and forest-clearings.
The usual nest-site is a hole in a tree, particularly abandoned holes made by Black Woodpeckers. It will take to using nest-boxes, and the provision of boxes will increase population-density in some areas. The species is very faithful to its nest-site from year to year.
Incubation is entirely by the ♀, who remains in the nesting hole for most of the breeding period, only leaving the nest for a short time in the early mornings. The ♂ provides food for both ♀ and young.
Tengmalm's Owls are mainly resident, though some dispersal takes place after the breeding season, particularly with northern populations.

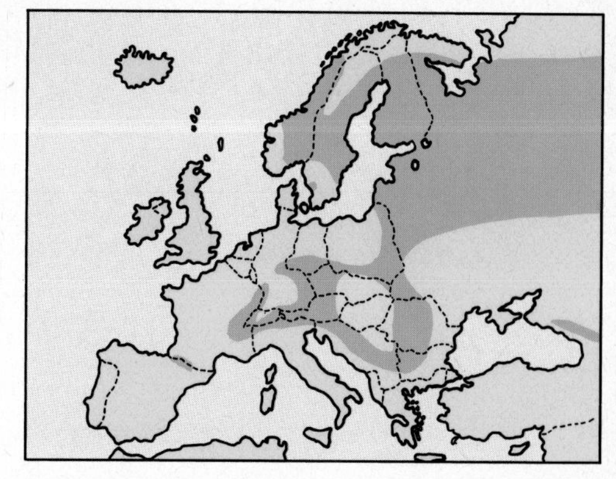

Length:	25 cm
Wing length:	♂ 16 cm, ♀ 18 cm
Weight:	♂ 130 g, ♀ 160 g
Voice:	'Bubububu', or a long-drawn-out trilling; also mewing and barking noises
Breeding period:	End of March, April, May, beginning of June. 1, exceptionally 2 broods per year. Replacement clutch may occur
Size of clutch:	4–6 (1–8) eggs
Colour of eggs:	White
Size of eggs:	32.6×26.5 mm
Incubation:	26–28 days, beginning from 1st egg
Fledging period:	Nidicolous; fledged at 31–34 days

, Baden-Württemberg, West Germany, 1968 (Schw)

Young bird, Baden-Württemberg (Becker)

Baden-Württemberg, 1968 (Schw)

) Ad., Finland (Hau)

Nightjar (Caprimulgus europaeus)

The Nightjar occurs in heathlands, open woodland with a dry understorey, sand-dunes, clearings in dense forest and arid areas. It feeds exclusively on large flying insects which are caught on the wing. The species is mainly crepuscular, hunting particularly at dusk. It has a buoyant flight, twisting and turning as it pursues its prey. It will also take insects from the ground that have been killed by road traffic. During the day it remains concealed, perching lengthways along branches or sitting on the ground, where its cryptic plumage affords good camouflage.
It has an acrobatic display which includes wing-clapping and hovering to show off the white wing- and tail-spots of the ♂. No nest is made; the eggs are merely laid in a hollow on the ground, often near some dead wood or other marker. Both sexes share in incubation, the ♀ usually sitting by day. The ♂ will take charge of the first brood while the ♀ begins a second clutch. Nightjars are wholly migratory, wintering in Africa south of the Sahara.
Migration: Depart breeding grounds in August–September, returning in April–May.

The larger, but similar **Red-necked Nightjar** (*Caprimulgus ruficollis*) breeds in the open sandy pine woods of Iberia. It winters in West Africa.

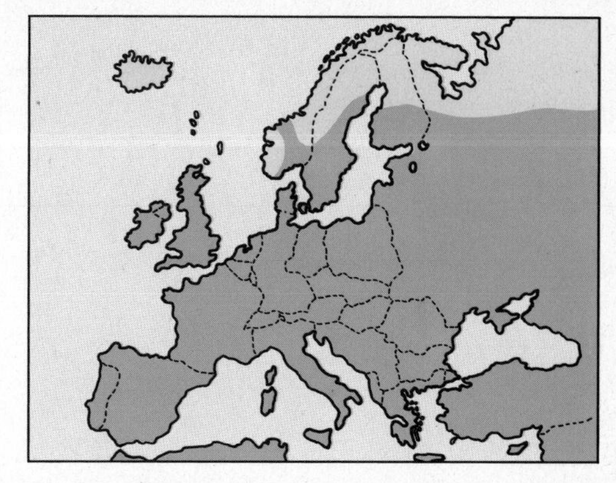

Length:	27 cm
Wing length:	18.5–20.2 cm
Weight:	75 g
Voice:	Courting call of the ♂ is a churring noise, rising and falling for some minutes. Flight-call: 'coo-ic'
Breeding period:	End of May, June, July, August and September. 2 broods per year. Replacement clutch possible
Size of clutch:	2 eggs
Colour of eggs:	Whitish, with grey-brown and grey-blue patches
Size of eggs:	31.9×22 mm
Incubation:	18–20 days, beginning from 1st egg
Fledging period:	Nidicolous; leaving the nesting place at 15 days, able to fly at 17–19, independent at about 30 days

ӧland, Sweden, July 1978 (Chr)

Juv., 15 days old, Lower Saxony, West Germany, 28.6.1977 (Sy)

Greece, 20.6.1976 (Li)

) Brooding ♀, Greece, 23.6.1976 (Li)

Swift (Apus Apus)

The original habitat of the Swift was rocky walls and cliffs. It is now found throughout towns and cities in Europe, seldom occurring away from habitation in the breeding season. It feeds entirely on insects which are caught on the wing. Large numbers of Swifts occur where food is abundant, often reaching a great height. Insects are stored in the throat when food is being collected for feeding the young. The Swift has a very rapid flight, sweeping round buildings and along streets at high speed, uttering the characteristic screaming call.
The nest-site is in a crevice or hole in a wall, cliff or nest-box; occasionally holes in trees are used. The nest is made of airborne plant material and feathers stuck together with saliva to form a shallow cup. Both parents share in incubation and care of the young. Swifts are entirely migratory, wintering throughout most of Africa south of the Sahara. They often make feeding movements in midsummer as they have to keep away from bad weather in order to obtain their airborne food.
Migration: Departs breeding areas in July–August with very few remaining into September. Return passage in April–June.

The similar **Pallid Swift** (*Apus pallidus*) breeds in Mediterranean countries. Its habits and food are very similar to those of the Swift and it also winters in Africa.

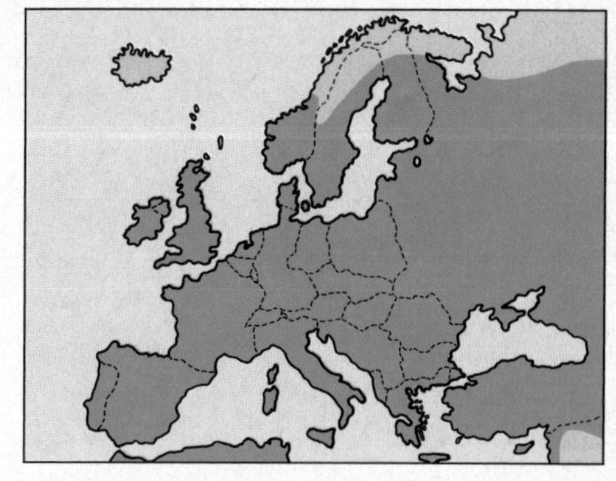

Length:	16.5 cm
Wing length:	16.5–18.0 cm
Weight:	40 g
Voice:	Shrill and penetrating 'sreeeh'
Breeding period:	Mid-May, June, 1 brood per year. Replacement clutch possible
Size of clutch:	3 (2–4) eggs
Colour of eggs:	White
Size of eggs:	25×16.3 mm
Incubation:	14–20 days, beg. from complete clutch
Fledging period:	Nidicolous; leaving the nest at about 6 weeks, and immediately independent

lged young bird, Bavaria, 1978 (Li)

Carinthia, 28.6.1978 (He)

Småland, Sweden, 30.6.1968 (Chr)

t) Carinthia, West Germany, 18.6.1978 (He)

Alpine Swift (Apus melba)

The Alpine Swift is found in rocky mountains and sea-cliffs in southern Europe. It also occurs in towns and cities, and in the north of its range this is the usual habitat.
Like the (Common) Swift, it feeds exclusively on flying insects which are caught on the wing. It is a large, powerful species with the rapid flight common to all swifts. It is often seen in flocks and it breeds colonially.
The nest-site is in caves, holes and rock-crevices (often in company with Rock Doves), or on ledges in buildings, towers, etc. The nest is constructed of feathers and plant material collected in the air and glued together with saliva to form a cup. The nest is gradually built up in successive years to form a substantial cup. Both parents share in incubation and in rearing the young. The species is wholly migratory, wintering in Africa south of the Sahara but north of the Equator.
Migration: Departs European breeding areas in October–November, returning in March and early April.

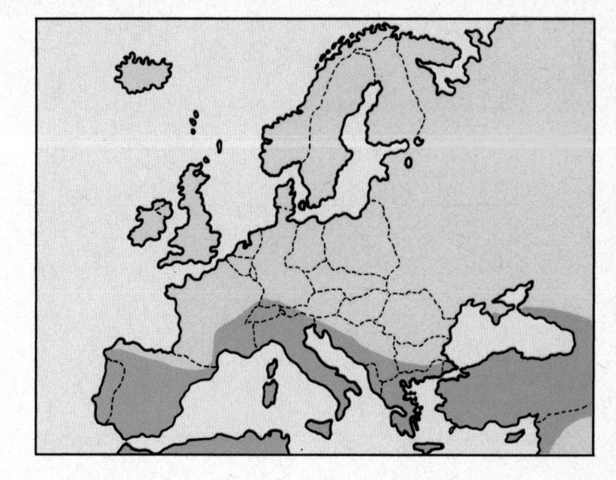

Length:	22 cm
Wing length:	20–23 cm
Weight:	100 g
Voice:	Generally mute. Flocks at breeding colonies make loud, chittering noise
Breeding period:	End of May, June. 1 brood per year. Replacement clutch possible
Size of clutch:	3 (1–4) eggs
Colour of eggs:	White
Size of eggs:	30.5 × 19.3 mm
Incubation:	18–23 days, beg. from complete clutch
Fledging period:	Nidicolous; leaving the nest and able to fly at *ca* 2 months

oding bird, Turkey, 15.6.1977 (Li)

') Ad., Turkey, 15.6.1977 (Li)

Ad. sheltering young, Carinthia, West Germany, 29.7.1977 (Zm)

Turkey, 15.6.1977 (Li)

Kingfisher (Alcedo atthis)

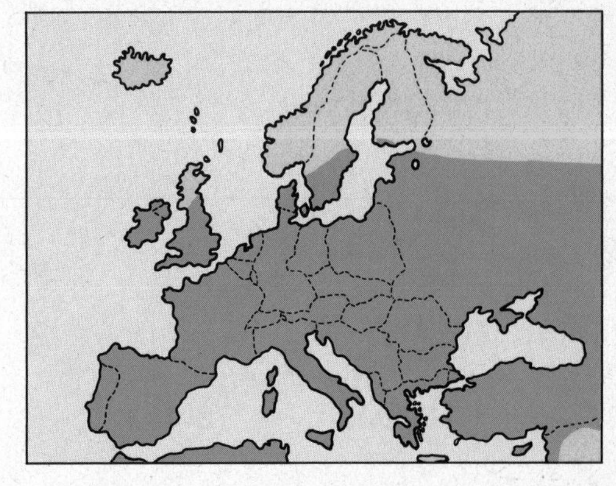

The Kingfisher is found near fresh water of most types including streams, rivers, canals, lakes and marshes. It prefers clear, slow-flowing waters. In winter it is often seen on tidal estuaries, salt marshes and rocky sea-shores. It feeds mainly on small fish such as Minnows, Sticklebacks and Gudgeon. It also takes crustaceans (an important part of the diet in winter), frogs, tadpoles and large insects. It catches its prey by sitting on an exposed perch overlooking the water and plunge-diving when prey is sighted. It will also hover for short periods above the water looking for prey. Kingfishers are highly territorial, especially during the breeding season, and are aggressive towards other Kingfishers that stray into the territory.
The nest-site is usually in a hole in the bank of a stream or lake, though sometimes it is a considerable distance from water. A tunnel 1–3 feet long is excavated, ending in a bowl-shaped chamber which contains the eggs. No nest material is used but the nest becomes lined with regurgitated fish-bones. Both sexes share in incubation and care of the young.
The Kingfisher is mainly resident and sedentary, though birds from the northern part of the range move southwards or to sea-coasts when fresh water freezes in winter. In hard winters there may be considerable movements.

Length:	16.5 cm
Wing length:	7.5–8.0 cm
Weight:	35 g
Voice:	A sharp whistle, repeated several times: 'chee' or 'chi-kee'
Breeding period:	End of April, May, June, July. Usually 2 broods per year. Replacement clutch possible
Size of clutch:	6–7 (4–8) eggs
Colour of eggs:	White
Size of eggs:	22.6 × 18.8 mm
Incubation:	19–21 days, beg. from complete clutch
Fledging period:	Nidicolous; leaving the nest at 4 weeks, and independent a few days later

♀, Hessen, September 1965 (Pf)

Fledged young, Amper, Bavaria, 1967 (Li)

Lower Saxony, West Germany, 26.4.1959 (Sy)

?) Ad. ♂, Hessen, West Germany, August 1967 (Pf)

Bee-eater (Merops apiaster)

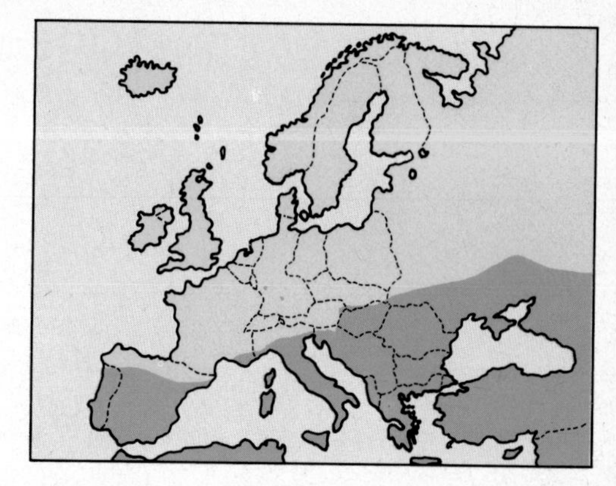

Bee-eaters occur mainly in warm Mediterranean countries, though sporadic breeding takes place well north of the normal range. It is found in a variety of habitats such as open country, woodland, scrub, arable land and plantations. It favours riverine situations where it can nest in the banks. It feeds in small groups, taking insects on the wing. It is often seen resting on wires, trees or hummocks. It has a graceful flight and is acrobatic in its pursuit of insects. During courtship the ♂♂ will present insects to the ♀♀. The species is always gregarious and nests in colonies. The nest is a tunnel up to 9 feet long, excavated by both parents in the sandy banks of a river, lake, cutting or steep slope. At the end of the tunnel is an unlined nest-chamber which holds the eggs. Incubation and care of the young is shared by both parents. 'Helpers' (unpaired birds) are known to assist with feeding at times.
Bee-eaters are entirely migratory, as there are insufficient insects present in southern Europe in winter. They winter in east and south Africa from the Equator southwards.
Migration: Departs breeding areas in August–September, returning in April–June.

Length:	28 cm
Wing length:	13.8–15.6 cm
Weight:	50 g
Voice:	'Pripripri' in series. Warning call: a sharp 'prit prit'
Breeding period:	Mid-May, June, July. 1 brood per year. Replacement clutch possible
Size of clutch:	4–7 (10) eggs
Colour of eggs:	White
Size of eggs:	25.6×21.7 mm
Incubation:	20–23 days, beginning from 1st egg
Fledging period:	Nidicolous; leaving the nesting tunnel and able to fly at about 4 or 5 weeks

at breeding place, Greece, June 1975 (Pf)

Lower Austria, July 1979 (Zm)

Excavated nest and clutch (note presence of insect-castings), Greece, June 1969 (Sy)

t) Courting pair, June 1971 (Li)

Roller (Coracias garrulus)

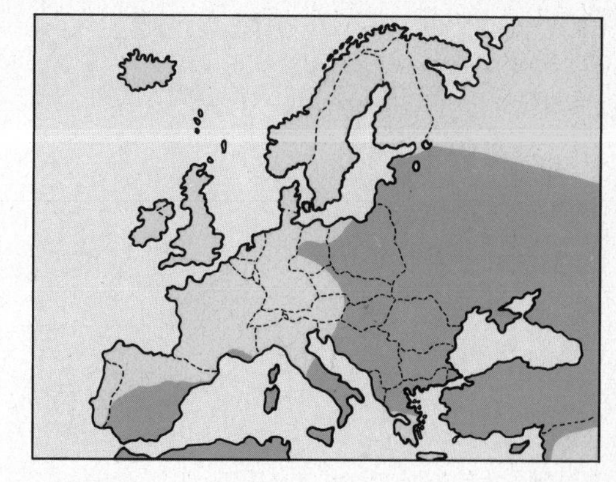

The Roller occurs in open country with scattered trees, open woodland, riverine areas and sunny mountain slopes. The diet is very varied. Insects make up the bulk, while small reptiles, scorpions, centipedes and even small birds or amphibians are taken at times. In winter, locusts are the main constituent. The food is taken from the ground, the birds often perching on wires or tree-tops and swooping down when prey is sighted. There is a remarkable display-flight, with the ♂♂ rising to a great height then tumbling and rolling over in a downward glide; hence the name. The nest-site is in a hole in a tree, bank of earth or in an old building or wall. Sometimes the abandoned nest of another bird is used. Often there is no nest-lining; at other times a sparse lining of plant material and feathers is furnished. Both sexes share in incubation and care of the young.
The range of the species in northern Europe has decreased due to climatic change. It required warm dry summers for successful nesting and is now mainly confined to southern and eastern Europe. Rollers are entirely migratory, wintering mainly in south and east Africa south of the Equator.
Migration: Departs breeding areas from August to September, returning in April–June.

Length:	31 cm
Wing length:	18.3–20.5 cm
Weight:	140 g
Voice:	'Rackrackrack' or 'rahrahrah'
Breeding period:	Mid-May, June. 1 brood per year. Replacement clutch possible
Size of clutch:	4–5 (6–7) eggs
Colour of eggs:	White
Size of eggs:	35.3×27.9 mm
Incubation:	18–19 days, beg. before clutch complete
Fledging period:	Nidicolous; leaving the nest at 26–28 days, independent a few days later

in front of nesting tunnel, Yugoslavia, June 1971 (Li)

Imm. sunbathing, Greece, 1976 (Li)

3- and 4-day-old chicks, Greece, 11.6.1976 (Li)

) Ad., Greece, June 1977 (Pf)

Hoopoe (Upupa epops)

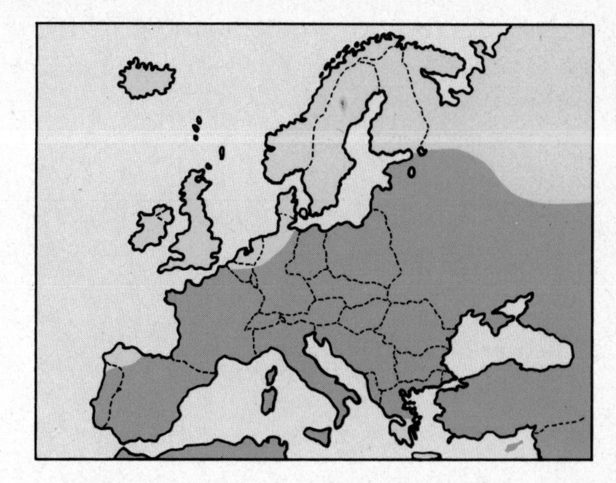

Hoopoes occur in a variety of dry habitats; open areas with scattered trees, orchards, large gardens, parkland, forest-clearings and riverine situations. It feeds chiefly on insect larvae, probing in soft ground with its long decurved bill. Other food items include small lizards, spiders, centipedes and grasshoppers. It has a varied choice of nest-sites, using holes in trees, in banks of earth, among rocks or in walls and buildings, including inhabited dwellings. Haystacks, nest-boxes and the abandoned burrows of small mammals are also used on occasion.

The nest itself is often unlined but at times plant material, scraps of cloth, feathers, wool, etc. may be used. The nest becomes fouled with droppings as the season progresses and may become evil-smelling. Incubation is by the ♀ alone, the ♂ providing food. He brings all food for the young in the early stages but both parents bring food when the young are well grown.

The Hoopoe is mainly migratory; a few winter in the extreme south of Europe but most move to winter in Africa in a belt south of the Sahara. The species occasionally breeds to the north of its normal range in Europe, but in general the northern populations are decreasing, probably due to climatic changes.

Migration: Departs European breeding areas in September–October, returning from late March to May.

Length:	28 cm
Wing length:	14.5–15.4 cm
Weight:	60 g
Voice:	Courting call: 'hoophoophoop', often repeated, and also hissing noises
Breeding period:	End of April, May, early June. 1 brood per year. Replacement clutch possible
Size of clutch:	5–8 (9–12) eggs
Colour of eggs:	Brownish to greenish grey, very variable
Size of eggs:	26×18 mm
Incubation:	16–18 days
Fledging period:	Nidicolous; leaving the nest at *ca* 26 days, still led by parents for a week or so

key, 20.5.1975 (Li)

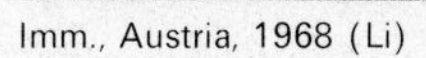

Imm., Austria, 1968 (Li)

Nest in building, Austria, 1968 (Rei)

) Yugoslavia, June 1971 (Li)

Green Woodpecker (Picus viridis)

The Green Woodpecker is found in broad-leaved and mixed woodland, parkland and orchards, and areas with scattered trees. It is normally found below 1,700 m though it is found at higher altitudes in Spain. It feeds chiefly on the larvae of wood-boring insects and ants. As a result it feeds a great deal on the ground and may be seen some distance from trees where ant-hills occur. It will also take fruit, mainly windfalls, and some seeds. It does not make a drumming noise but frequently make a loud tapping noise when searching for insects on tree-trunks.
The nest-site is a hole in a tree, excavated by both sexes, usually at least 1 m above the ground. Some holes may be excavated for roosting. The nest-chamber is unlined. Both sexes share in incubation and care of the young. The young are fed on a regurgitated semi-liquid paste.
Because the Green Woodpecker does not live in forests and dense woodland it is subject to severe food-shortage in hard winters. This often results in a considerable decrease in numbers, particularly in the northern populations. The species is resident and sedentary throughout its range.

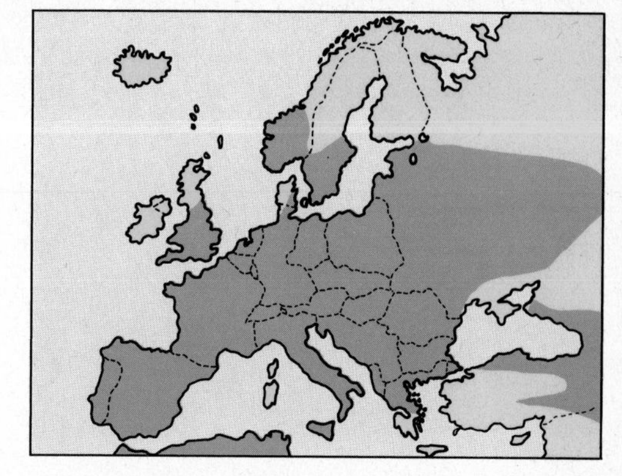

Length:	32 cm
Wing length:	15.7–16.7 cm
Weight:	200 g
Voice:	♂ and ♀ have a ringing laugh, 'kyuckyuckyuck'; drumming is exceptional
Breeding period:	End of April, May. 1 brood per year. Replacement clutch possible
Size of clutch:	6–7 (4–9) eggs
Colour of eggs:	White
Size of eggs:	31×23 mm
Incubation:	18–19 days, beg. from complete clutch
Fledging period:	Nidicolous; leave the hole at 3 weeks and are tended a further 3 weeks

nesting hole, Bavaria (Li)

Imm. seeking food, Württemberg (Schu)

') ♂ feeding at ant-hill, Bavaria (Wo)

Grey-headed Woodpecker

(Picus canus)

The Grey-headed Woodpecker inhabits light broad-leaved and mixed woodland in montane areas as well as parkland, open areas with scattered trees and deciduous forest in lowland areas. It occurs at a much higher altitude than the Green Woodpecker and will live in denser woodland with more conifers than that species. It feeds on wood-boring insects and their larvae, a wide variety of other insects, ants, seeds and artificially provided food such as fat. It spends less time ground-feeding than the Green Woodpecker.

The nest-site is a hole in a tree, usually a deciduous tree and often well into the forest. The hole is normally between 1 and 15 m from the ground and is excavated by both parents. The nest-cavity at the end of the tunnel is lined with wood-chips. Incubation and rearing the young is shared by both parents. The young are fed on regurgitated matter.

The Grey-headed Woodpecker is slowly increasing its range through western Europe, occupying areas traditionally held by the Green Woodpecker.

The Grey-headed Woodpecker is resident and sedentary other than the gradual range-extension.

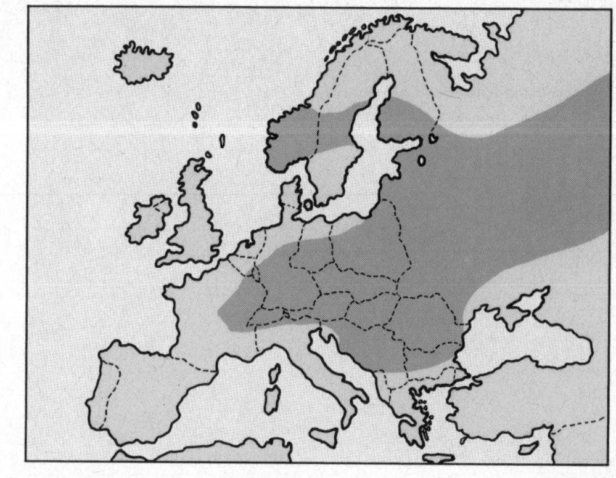

Length:	26 cm
Wing length:	14.3–15 cm
Weight:	130 g
Voice:	♂ and ♀ a laughing 'gyukyukyuk', slower and more musical than Green Woodpecker, drums more frequently than that species
Breeding period:	March–June. 1 brood per year
Size of clutch:	6–7 (5–9) eggs
Colour of eggs:	White
Size of eggs:	29.5×22.8 mm
Incubation:	17–18 days, beg. from complete clutch
Fledging period:	Nidicolous; young leave nest after *ca* 3 weeks

front of nesting-hole, Bavaria, June 1977 (Pf)

Bavaria, 3.2.1963 (Br)

♂ feeding young bird, Bavaria (Diep)

) ♂ in front of nesting-hole, Bavaria, June 1978 (Pf)

Black Woodpecker (Dryocopus martius)

The Black Woodpecker is the largest of the European species. It is found in coniferous, mixed and broad-leaved forest (particularly beech) with tall, mature trees. It frequently occurs in montane situations as high as 2,000 m. It feeds on wood-boring insects and their larvae, chipping into the bark and wood with its powerful bill. It also takes a large number of the big forest ants. In winter it will dig into the snow to reach tree-stumps or rotten wood where the ants are found.

It nests in a hole in a tree, usually beech or pine. Both sexes excavate the tunnel, which is normally 3–10 m above the ground. The nest-cavity is unlined or has a layer of wood-chips. A new nest-site is chosen each year, the old holes being used for roosting. These holes are also used as nest-sites by the smaller owls and other birds. Both sexes share in incubation and care of the young, The ♂ takes the larger share and usually sits at night. The young are fed by regurgitation.

Black Woodpeckers are mainly resident and sedentary, though some birds from the north of the range may disperse in winter. The species is slowly extending its range into western Europe, though there is some evidence that there is competition from Green and Grey-headed Woodpeckers.

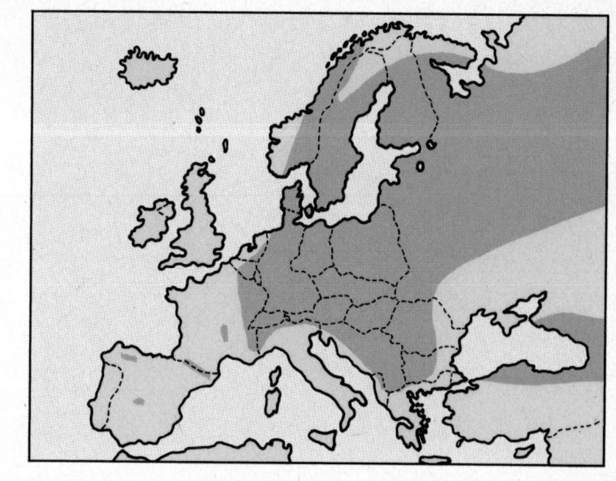

Length:	46 cm
Wing length:	23.5–25.8 cm
Weight:	300 g
Voice:	'Grree grree' and 'klee-oh', a ringing 'quikquikquik'
Breeding period:	Mid-April to May. 1 brood per year. Replacement clutch possible
Size of clutch:	4–5 (3–6) eggs
Colour of eggs:	White
Size of eggs:	34×24 mm
Incubation:	12–14 days, beg. after laying of 1st egg
Fledging period:	Nidicolous; fledged at about 4 weeks, but led a further 1–2 months

. front of nesting hole, Bavaria, April 1974 (Pf)

Juv. ♂, Bavaria, 1975 (Li)

Lower Saxony, 27.4.1958 (Sy)

') Ad. and juv. ♂, Bavaria, 1975 (Li)

Great Spotted Woodpecker

(Dendrocopos major)

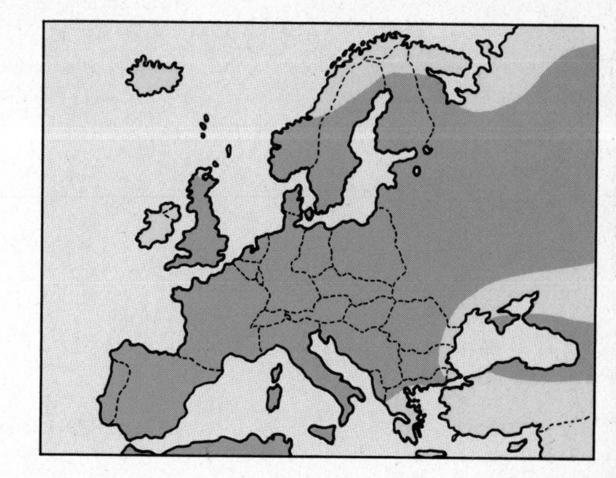

The Great Spotted Woodpecker is the most widespread of the Palearctic woodpeckers. It is found in most types of woodland, large gardens, parks, riverine situations and scattered trees in open country. It is found in mountainous areas as high as the tree-line. The diet is a variety of wood-boring insects and larvae, spiders, nuts, seeds and berries. It will come to artificially provided food such as fat and has a habit of wedging pine-cones into a crevice, known as a 'smithy', in order to chip out the seeds. The nest is in a hole in a tree, usually 2 m or more above ground. Both sexes excavate the hole, which may be in a rotten or sound tree, either deciduous or conifer. The ♀ does most of the incubation, with both parents rearing the young. The young are fed on insects which are carried in the parent's bill.

The species makes a characteristic drumming noise, made by hammering the bill against the end of a branch or a similar resonant surface. This drumming is often audible up to 400 m and is frequently the first sign of its presence in a wood.

The species is mainly resident and sedentary, though northern populations will make southward dispersive movements when food is scarce. Often these movements take the form of an 'invasion', with large numbers of woodpeckers arriving at coastal sites after a North Sea crossing.

Length:	23 cm
Wing length:	12.6–13.8 cm
Weight:	80 g
Voice:	Typical 'tchick', deeper and louder than the Lesser Spotted Woodpecker. ♂ and ♀ drum frequently at courting time
Breeding period:	Mid-May to June. 1 brood per year. Replacement clutch possible
Size of clutch:	5–7 (4–8) eggs
Colour of eggs:	White
Size of eggs:	26×20 mm
Incubation:	About 16 days, beg. from complete clutch
Fledging period:	Nidicolous; leaving hole at 3 weeks and remaining a further 2 weeks under protection of parents

♂, Bavaria, 15.1.1963 (Br)

Young in hole, Bavaria, 14.6.1972 (Li)

Jutland, Denmark, 17.6.1955 (Chr)

(Left) Ad. ♂ at 'smithy', Bavaria, 15.12.1979 (Li)

Syrian Woodpecker

(Dendrocopos syriacus)

The Syrian Woodpecker has spread rapidly in Europe since it colonised from Asia Minor in the late 1800s. It is now found over much of lowland south-east Europe. It occurs in open, cultivated land with scattered trees, light deciduous woodland, riverine trees, parks, gardens and in towns. It is not found in mountainous areas, where the Great Spotted Woodpecker is present, and generally avoids conifers.
It feeds on a variety of insects, spiders, seeds and fruit, the latter forming a significant part of the diet.
The nest-site is a hole in a tree, particularly cherry, walnut and mulberry trees. The tunnel is excavated by both sexes and the nest-chamber is unlined or has a layer of wood-chips. Incubation and care of the young is shared by both parents. The ♂ incubates by night, the ♀ by day. The young are fed on insects and fruit carried in the parent's bill.
The species is resident and sedentary apart from the gradual westward extension of the range.
Some hybrids with Great Spotted Woodpecker have been recorded, particularly on the edge of the range. The two species are very similar in appearance and habits.

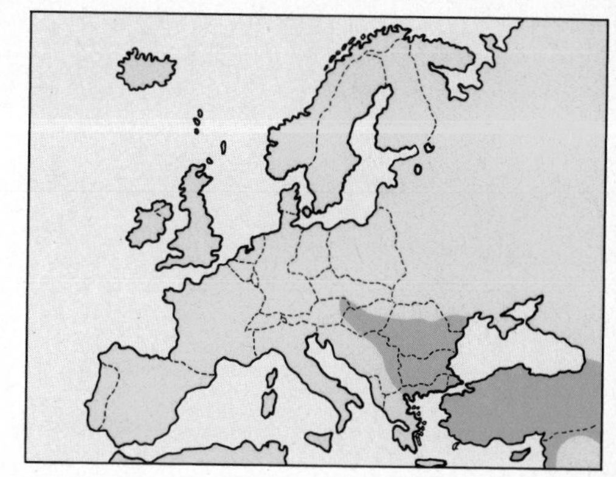

Length:	23 cm
Wing length:	12.5–13 cm
Weight:	80 g
Voice:	Like that of the Great Spotted Woodpecker, but softer: 'chiq', and a Moorhen-like 'kirrook'
Breeding period:	Early May to early June. 1 brood per year. Replacement clutch possible
Size of clutch:	5 (4–7) eggs
Colour of eggs:	White, indistinguishable from those of Great Spotted Woodpecker
Size of eggs:	26×18.7 mm
Incubation:	11–12 days, beg. from complete clutch
Fledging period:	Nidicolous; fledged at 24 days, independent after a further 2 weeks

ireece, 1.7.1973 (Sy)

Juv., Vienna, 25.5.1972 (Fe)

Prepared exhibit for a zoological collection (Pf)

) ♂ with food, Austria (Mal)

Middle Spotted Woodpecker

(Dendrocopos medius)

The Middle Spotted Woodpecker is found in broad-leaved forests with a preponderance of oak and hornbeam trees. It also occurs in parks, old orchards and gardens. due to the decrease in deciduous woodland in Europe, the species is now rather scarce with a very patchy distribution. It feeds mainly on small insects, which it picks out of cracks in the bark of trees. It does not hammer at bark to find food like the Great Spotted Woodpecker. It will also hunt for food among the leaves in the tops of trees and will come to bird-tables and artificial feeding-sites in winter.

The nest-site is a hole in a tree, almost always a deciduous tree. Both sexes excavate the hole, though the ♂ takes the larger share, and the hole is usually about 4 m above the ground. Occasionally a natural hole is used. The nest-cavity is unlined. Both sexes share in incubation and care of the young.

The Middle Spotted Woodpecker is largely resident and sedentary throughout its range.

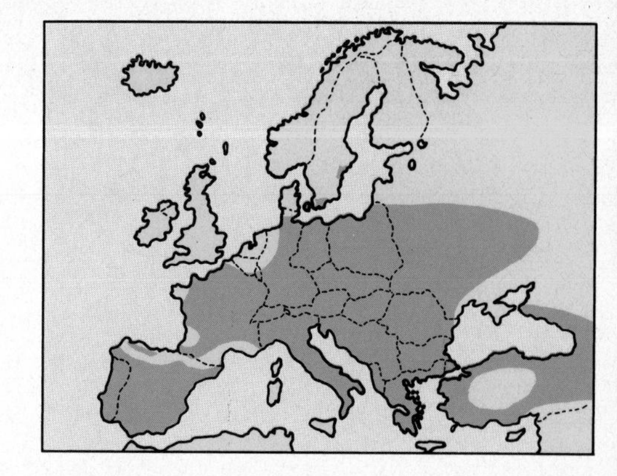

Length:	21 cm
Wing length:	12.3–12.9 cm
Weight:	60 g
Voice:	Rarely drums. For both sexes, characteristic call is a nasal 'wait-wait'
Breeding period:	End of April, May, early June. 1 brood per year. Replacement clutch may occur
Size of clutch:	5–6 (4–8) eggs
Colour of eggs:	White
Size of eggs:	23×18 mm
Incubation:	12–13 days, beg. from complete clutch
Fledging period:	Nidicolous; remaining 3 weeks in the hole and a further 8–10 days under parental protection

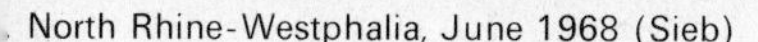

. North Rhine-Westphalia, June 1968 (Sieb)

Ad., Lower Saxony, January 1972 (Di)

Imm., Württemberg, June 1974 (Schu)

) Ad. feeding young, North Rhine-Westphalia, June 1968 (Sieb)

Three-toed Woodpecker

(Picoides tridactylus)

This woodpecker has two separate populations in Europe; one in the coniferous woodland on the mountains in central and south-east Europe, the other in the mixed coniferous and birch forest of the north. It feeds on insects and their larvae on lichen-covered trees. Rarely hammering with the bill, it delicately prises out the insect from a crevice. The nest-site is normally in a conifer, the hole being excavated by both parents. Incubation and care of the young is shared and the young are fed on insects carried in the bill of the parent. The species is resident and sedentary.

White-backed Woodpecker (*Dendrocopus leucotos*) This is the most scarce of European woodpeckers with a highly disjunct distribution. It favours broad-leaved woodland, particularly beech, riverine aspen woods and mixed conifer/birch woodland. It requires a habitat with many dead, ancient trees. It feeds on large woodboring insects and larvae which are obtained by hammering with the strong bill. The nest-site is in a hole in a tree, usually a rotten deciduous trunk. The incubation and care of the young is shared by both sexes. It is resident and sedentary throughout its range.

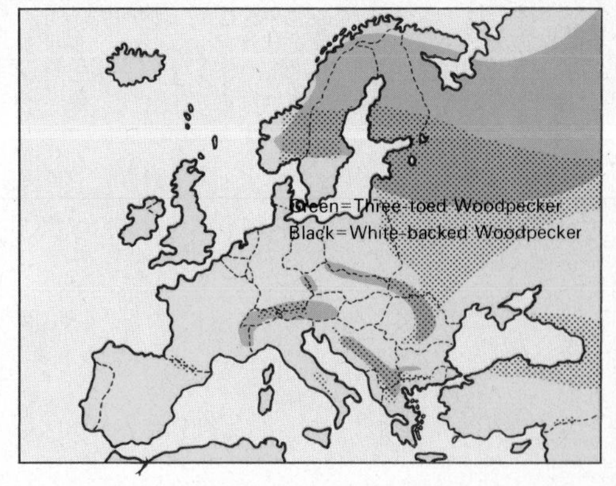

Length:	22 cm
Wing length:	12–12.9 cm
Weight:	70 g
Voice:	'Kek-ek-ek', like Great Spotted Woodpecker but deeper. Both sexes drum slowly
Breeding period:	End of May to June. 1 brood per year
Size of clutch:	3–5 eggs
Colour of eggs:	White
Size of eggs:	24×18 mm
Incubation:	14 days
Fledging period:	Nidicolous; leaving the hole at 3 weeks, and led a further 5–6 weeks

e-toed Woodpecker ♂ (Hau)

White-backed Woodpecker ♂ (Hau)

White-backed Woodpecker ♀, Austria 1974 (Schw)

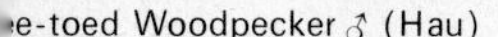

) Three-toed Woodpecker ♀, Austria, 10.6.1977 (Zm)

Lesser Spotted Woodpecker

(Dendrocopos minor)

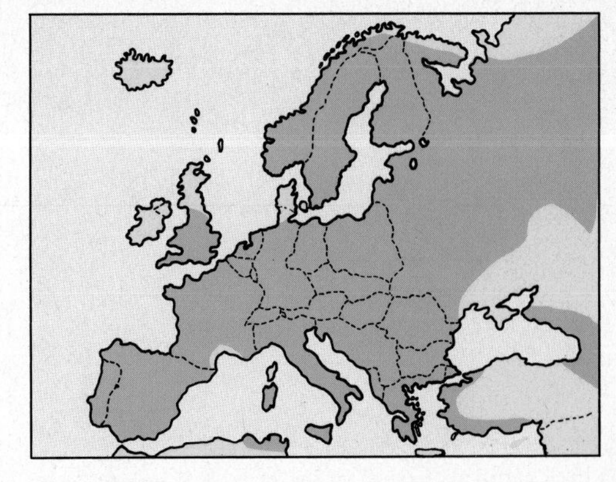

This, the smallest of the European woodpeckers, inhabits open woodland – both deciduous and mixed – parks, orchards, gardens and other areas with scattered trees. It feeds on a variety of insects including the woodboring types and their larvae. It prefers to forage on side-branches and twigs to the main trunk. Some berries and seeds are taken and the species wil visit bird-tables and other artificial feeding-sites in winter.
The nest-site is a hole in a tree, usually in decayed wood in a branch or stump. The hole is excavated by both sexes and the nest-chamber is unlined save for a few wood-chips. Incubation is by both sexes, the male sitting at night. The young are fed on insects brought in the bill of the parents.
The species is mainly resident and sedentary throughout its range. Occasional dispersal of northern populations is known when food becomes scarce in winter.

Length:	15 cm
Wing length:	8.5–9.6 cm
Weight:	22 g
Voice:	♂ and ♀ generally utter series of calls, 'kikikiki', like that of Wryneck; both drum frequently at breeding time
Breeding period:	May–June. 1 brood per year. Replacement clutch possible
Size of clutch:	5–6 (3–8) eggs
Colour of eggs:	White
Size of eggs:	19×14 mm
Incubation:	14 days, beg. from complete clutch
Fledging period:	Nidicolous: remaining 3 weeks in nest, and led a further 2 weeks after fledging

♀, Bavaria, (Diep

♂, Lower Saxony, June 1979 (Sy)

♂, feeding young, Carinthia, West Germany (Zm)

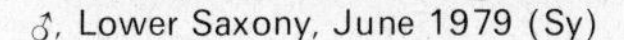

♂ in front of nesting hole, Bavaria, June 1979 (Pf)

Wryneck (Jynx torquilla)

The Wryneck is found in a variety of habitats. In the north of the range it occurs in larch and pine woods, in the south if frequents open deciduous woodland, orchards, parkland with scattered trees, gardens and marginal woodland along river valleys and roadsides. It feeds mainly on ants and larvae and a variety of insects. It will search for prey on the ground as well as on the trunk and branches of trees. It does not hammer or bore for insects like some woodpeckers. The nest-site is normally a hole in a tree, a nest-box or occasionally a hole in a building or bank. They do not excavate their nests but take over existing holes. The nest itself is normally unlined or on existing wood-chips. Incubation is shared but is mainly done by the ♀. Both sexes tend the young.
The species has decreased considerably in England and the Channel coast of Europe but in recent years parts of Scotland have been colonised by birds of the Scandinavian population which breed in pine trees. The species is almost entirely migratory, wintering in Africa from the Sahara southwards to the Equator. Some birds overwinter in the extreme south of Europe or North Africa.
Migration: Departs breeding grounds in August–September, returning from late March to May.

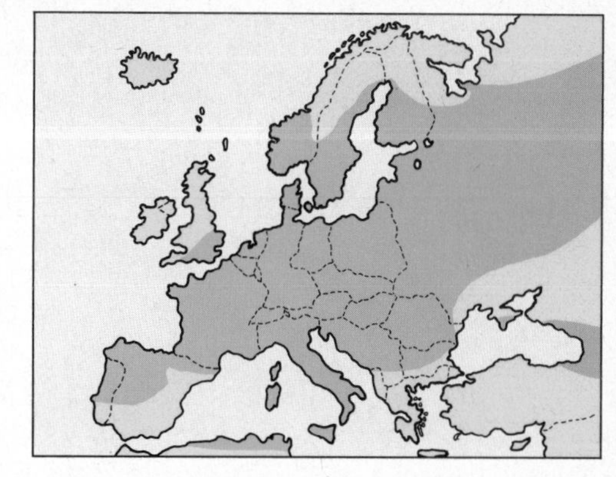

Length:	16.5 cm
Wing length:	8.3–9.2 cm
Weight:	35 g
Voice:	Courting call like Lesser Spotted Woodpecker, 'quee-quee-quee-quee-quee'
Breeding period:	End of May, June. 1 or sometimes 2 broods per year
Size of clutch:	7–10 (5–14) eggs
Colour of eggs:	White
Size of eggs:	20.4×15.5 mm
Incubation:	13–14 days, beg. from complete clutch
Fledging period:	Nidicolous; leaving the hole at 19–22 days, and tended a further 10 days by the parents

aria, 1966 (Li)

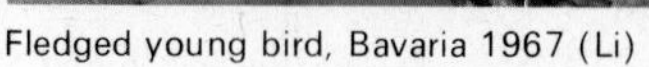

Fledged young bird, Bavaria 1967 (Li)

Småland, Sweden, 17.6.1973 (Chr)

) Ad., Austria, 2.6.1960 (Fe)

Lesser Short-toed Lark

(Calandrella rufescens)

This species forms a sibling with the Short-toed Lark with which it shares part of its range. In Europe it is found in arid areas, salt marshes, salicornia scrub, stony semi-desert and barren steppe. It feeds on small seeds and green shoots; some insects are also taken. Outside the breeding season the species occurs in flocks, sometimes of considerable size where food is plentiful.
The ♂ has a display-flight, rising steeply in a spiral or a high circling flight. The nest-site is on the ground, usually in the shelter of some thorn-scrub or other vegetation. The nest is a deep cup, built in a hollow and constructed of grass, stems and other plant material. Both sexes share in incubation and care of the young, the ♀ taking the larger part.
The European population is migratory, wintering in North Africa and parts of the Middle East, where it occurs alongside the resident population of those areas.

Migration: Departs breeding areas in September–October, returning in March and April.

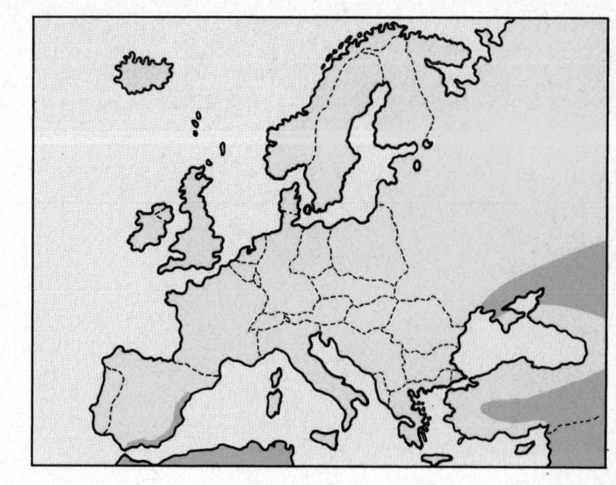

Length:	14 cm
Wing length:	8.5–9.3 cm
Weight:	20 g
Voice:	'Prritt'; song resembles that of Short-toed Lark, but longer and more musical; it imitates other birds
Breeding period:	Early April, May, June. 2 broods per year. Replacement clutch possible
Size of clutch:	3–4 (5) eggs
Colour of eggs:	Off white, pale mud-brown and ash-grey patches, often concentrated at the blunt pole
Size of eggs:	20×14.8 mm
Incubation:	13 days, beg. from complete clutch
Fledging period:	Nidicolous; leaving the nest at 10 days, not yet able to fly

. brooding, Turkey, 28.5.1974 (Sy)

Tuz-Gölü, Turkey, 27.5.1974 (Li)

t) Ad., Tuz-Gölü, Turkey, 28.5.1974 (Sy)

Short-toed Lark

(Calandrella brachydactyla)

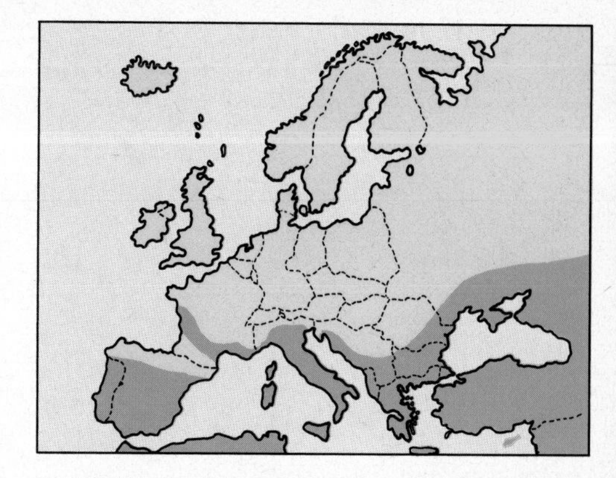

The Short-toed Lark inhabits open dry country such as semi-desert, steppe, stony agricultural land, sand-dunes and dry salt flats. It feeds on a variety of seeds, green shoots and small insects. The ♂ has a distinctive display-flight, rising and falling slowly in the manner of a yo-yo. It sings persistently throughout the display.
The nest-site is on the ground, usually in the shelter of a tuft of vegetation; sometimes the species nests in loose colonies. The nest is a deep cup of grasses, stems and other plant material, lined with wool, hair or feathers built in a natural hollow. Both sexes share incubation, the ♀ taking the major part. The young are fed by both parents.
The Short-toed Lark is entirely migratory, wintering in a narrow belt of arid country south of the Sahara. Outside the breeding season the species forms large flocks.
Migration: Departs breeding areas from September to November, returning in late March to early May. Sometimes overshoots breeding areas and occurs as a vagrant in northern Europe.

Length:	14 cm
Wing length:	8.5–9.6 cm
Weight:	22 g
Voice:	Call: 'tchi-tchirrup'; also short phrases, not unlike Willow Warbler
Breeding period:	End of April, May, June. 2 broods per year. Replacement clutch possible
Size of clutch:	4 (3–5) eggs
Colour of eggs:	Yellowish white with fine light or olive-brown patches, generally concentrated at the blunt pole
Size of eggs:	20×14.6 mm
Incubation:	13 days, beg. from complete clutch
Fledging period:	Nidicolous; leaving the nest at 9–12 days, and able to fly within a few days

ng birds, Tuz-Gölü, Turkey, 1970 (Wa)

Macedonia, Greece, June 1977 (Pf)

Tuz-Gölü, Turkey, 28.5.1974 (Li)

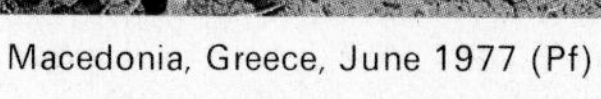

) Ad., Macedonia, Greece, 2.6.1976 (Pf)

Calandra Lark (Melanocorypha calandra)

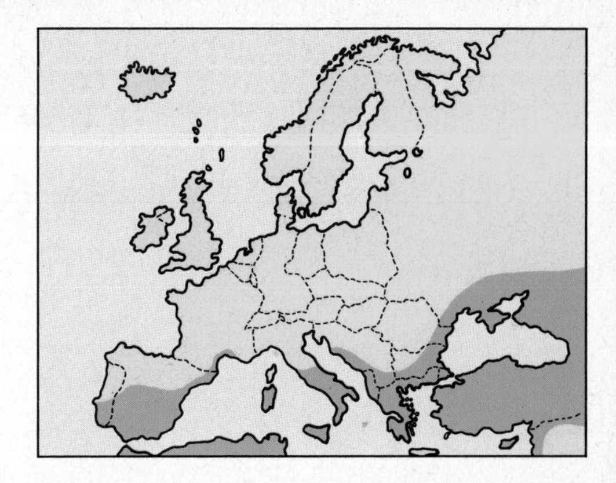

The Calandra Lark is the largest of the European larks. It occurs in grassy steppes, open pastures, dry agricultural land and arid brush. It feeds on seeds, shoots of young plants and insects, particularly grasshoppers.
The ♂ has a high, circling display-flight often dropping silently to the ground at the end. The flight is buoyant with a characteristic slow wing-action.
The nest-site is on the ground, usually amongst thistles or other prickly plants rejected by grazing animals. The nest is a cup of grasses and stems, built into an existing hollow. The nest is built by the ♀ and she incubates alone, the ♂ standing guard nearby. Both sexes being food for the young.
Outside the breeding season the species is gregarious, often occurring in large flocks where food is abundant (stubble-fields, etc.). The Calandra Lark is mainly resident in west and southern Europe, but eastern birds migrate to winter in the Middle East and part of north-east Africa.

Length:	19 cm
Wing length:	12.7–13.4 cm
Weight:	65 g
Voice:	♂ sings sitting and in circular flight, like Skylark, and often mimicking other birds. Call: jangling 'kleetra'
Breeding period:	April to mid-June. 2 broods per year. Replacement clutch possible
Size of clutch:	4–5 (–7) eggs
Colour of eggs:	Whitish to yellow-green background, with thick patches in various shades of brown
Size of eggs:	24 × 18 mm
Incubation:	*ca* 16 days, beg. from complete clutch
Fledging period:	Nidicolous; leaving the nest unable to fly at 10 days, fledged a few days later

♂, Turkey, 10.6.1977 (Li)

Fledged young bird, Turkey, 25.6.1977 (Li)

Tuz-Gölü, Turkey, 26.5.1974 (Li)

) Ad., Turkey, May 1974 (Pf)

Shore Lark (Eremophila alpestris)

The Shore Lark has a very discontinuous distribution. It is mainly a bird of the high mountainous areas, usually above 4,000 feet and did not colonise Scandinavia, where it nests on the fells at much lower altitudes, till the last century. It is found in alpine meadows, rocky areas up to the snow-line and in the tundra of the far north. In winter it may be found on salt marshes, dunes and beaches. It feeds on a variety of small insects, seeds, buds and shoots of dwarf vegetation. In winter small molluscs and crustaceans are taken.
The song is brief and usually uttered from a perch on the ground. Though it has a display-flight similar to that of the Skylark it does not sing for long periods in flight.
The nest is sited on the ground usually in the shelter of alpine plants or stones. The nest is a loose cup of grass and other plant material, sometimes lined with hair or plant 'down'. Sometimes a rim of small pebbles or peat fragments is placed round the nest. Incubation is by the ♀ alone; both sexes share in rearing the young. In winter it is gregarious and is often met with in small flocks. The northern population is entirely migratory, wintering in the southern Baltic, Channel coasts and the east coast of England. The south-east population is generally resident, though some altitudinal movements take place in winter.
Migration: Departs breeding areas in October–November, returning in March–May.

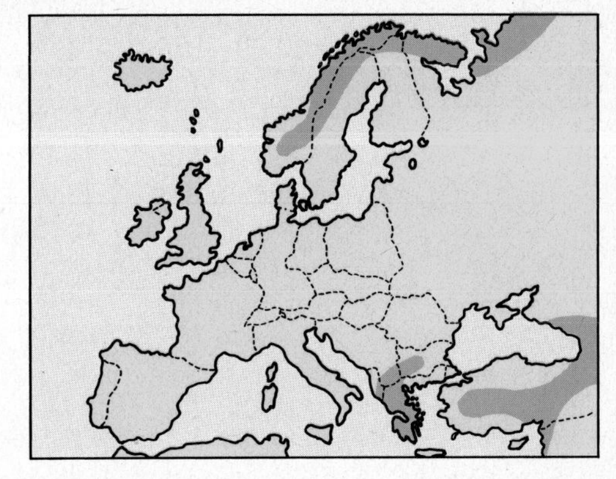

Length:	17 cm
Wing length:	10.6–11.2 cm
Weight:	40 g
Voice:	Pipit-like 'tsee-tsi'; tinkling song-flight like weak Skylark
Breeding period:	Mid-May, June, Mid-July. 2 broods per year
Size of clutch:	4–5 (3–6) eggs
Colour of eggs:	Greenish white, sprinkled with more or less reddish brown
Size of eggs:	23×16 mm
Incubation:	10–14 days, beg. from complete clutch
Fledging period:	Nidicolous; leaves the nest at 9–12 days, fledged soon after

♀, Fokstua, Norway, 26.6.1974 (Pl)

Ad. ♂ (*E.a. penicillata*) Turkey, 10.6.1977 (Li)

Turkey, 9.6.1977 (Li)

t) Ad. ♂, Fokstua, Norway, 29.6.1974 (Pl)

Woodlark (Lullula arborea)

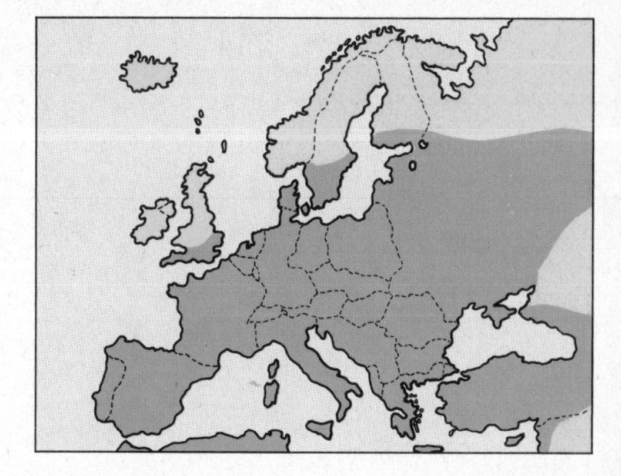

Throughout its range the Woodlark is a rather local bird, perhaps due to the type of habitat requirements. It is found in dry, sandy heathland with scattered trees, the edges of woodland bordering heaths, orchards, olive groves and alpine meadows with juniper scrub. The presence of a light sandy soil and scattered trees is an important factor.
It feeds on a wide variety of small insects and spiders; in winter, seeds are a major constituent of the diet.
Outside the breeding season it occurs in small parties but is not very gregarious and rarely forms large flocks.
The nest is on the ground, usually well hidden in vegetation and very difficult to find. It is built of grasses and stems, mainly by the ♂, and sometimes lined with hair or fine grass by the ♀. Incubation is by the ♀ alone, with both sexes rearing the young. The ♂ has a display-flight which involves circling the territory, often at a considerable height, with a fluttering action. The bird sings throughout this flight then drops to the ground or a tree with folded wings. It is often heard singing at night.
The species is resident in western Europe but the eastern and Scandinavian populations migrate south-west to winter in southern Europe and parts of North Africa.
Migration: Departs breeding areas in October to late November, returning in February–March.

Length:	15 cm
Wing length:	9.2–9.9 cm
Weight:	20–30 g
Voice:	Typical song of ♂: a repetition of 'tu tu tu', fluty and musical
Breeding period:	End of March to June, in northern Europe end of April to July. 2–3 broods per year
Size of clutch:	4–5 (3–6) eggs
Colour of eggs:	Whitish, with more or less dense brown or red-brown spots, usually crowning the blunt pole
Size of eggs:	21×16 mm
Incubation:	*ca* 14 days, beg. from complete clutch
Fledging period:	Nidicolous; leaving the nest at 12 days, able to fly shortly afterwards

Bavaria, June 1977 (Wo)

Nearly fledged young, Greece, 3.6.1978 (Li)

Lower Saxony, 12.5.1979 (Sy)

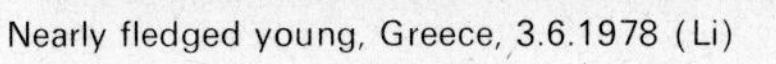

*) Ad. feeding young, Lower Saxony, 23.4.1978 (Sy)

Crested Lark (Galerida cristata)

The Crested Lark is found in dry open areas of agricultural land, rocky or sandy areas, waste ground, roadsides and similar areas with low vegetation. It is more tolerant of trees than the Skylark and often perches on trees, walls and buildings.
It feeds on a variety of small seeds, shoots, insects and spiders. It is not particularly gregarious and is rarely seen in flocks. It is often very tame and tolerant of man.
The song is short and less musical than that of the Skylark and is usually uttered from the ground or a vantage-point. The song-flight is poorly developed.
The nest-site is in a hollow on the ground, often in crops or waste ground. Sometimes it will nest on flat roofs. The nest is a loose cup of grass and other plant material built by both sexes, often built above ground-level. Incubation is mainly by the ♀ with both sexes rearing the young.
The Crested Lark is resident and mainly sedentary throughout its range.
In Iberia the similar **Thekla Lark** (*Galerida theklae*) occurs. It prefers more hilly and rocky habitats but is otherwise very similar in both habits and appearance to the Crested Lark. The song is different and the display-flight more developed.

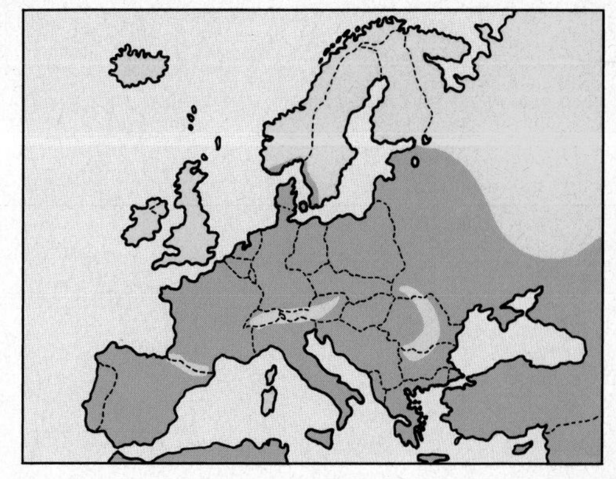

Length:	17 cm
Wing length:	9.7–10.9 cm
Weight:	45 g
Voice:	'Whee-wheeoo' and 'dui'; several series of notes with variants
Breeding period:	Mid-April to mid-July. 2–3 broods per year
Size of clutch:	4–5 (3–6) eggs
Colour of eggs:	Whitish grey with yellow-brown patches, often forming a crown at the blunt pole
Size of eggs:	22.8×16.8 mm
Incubation:	12–13 days, beg. from complete clutch
Fledging period:	Nidicolous; concealing themselves while still unable to fly near the nest; independent at 15 days

'. ♂ at nest (*G.c. meridionalis*), Greece (Sy)

Turkey, Tuz-Gölü, 8.5.1975 (Li)

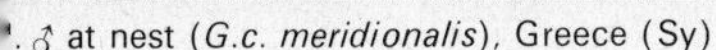

ft) Yugoslavia, 1971 (Li)

Skylark (Alauda arvensis)

The Skylark is widely distributed in Europe and occupies a range of habitat from cultivated land to moor, heath, marshes, alpine meadows, steppe and other open areas. It generally avoids wooded country. It has extended its range considerably due to increased agricultural development providing suitable habitat.

It feeds on seeds, green shoots, insects, spiders, small worms and other invertebrates. It is gregarious outside the breeding season, often occurring in large flocks.

The song is pleasant and musical, delivered in a sustained hovering song-flight, often at a great height.

The nest is in a hollow in the ground, often in the cover of some low vegetation, and is constructed of grasses and other plant material by the ♀. She incubates alone but both sexes share in rearing the young.

Though resident in the south and west of the range, the more northerly and eastern populations are migratory. They winter mainly in western Europe, parts of North Africa and the Middle East.

Migration: Mainly September–October but hard-weather movements occur throughout the winter months. Spring passage in February–April.

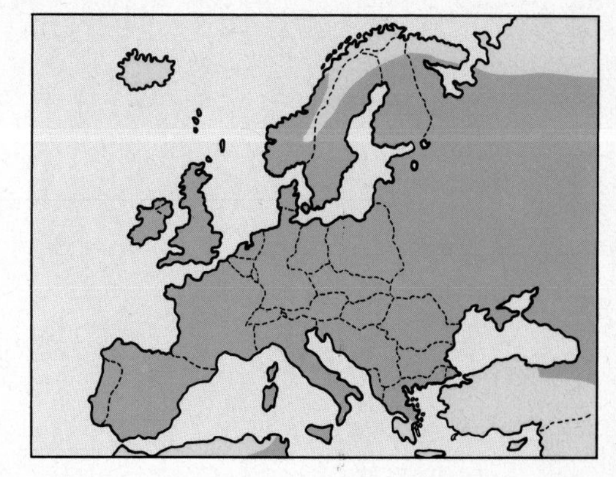

Length:	18 cm
Wing length:	9.8–11.5 cm
Weight:	40 g
Voice:	A jingling 'chirrup'. Song: a loud musical warble
Breeding period:	Mid-April to July. 2–3 broods per year
Size of clutch:	4 (3–5) eggs
Colour of eggs:	Whitish to cream, thickly spotted with large or small brownish patches, often in a ring at the blunt pole
Size of eggs:	23×17 mm
Incubation:	11–14 days, beg. from complete clutch
Fledging period:	Nidicolous; leaving the nest at 10 days, unable to fly; flies well after a further 10 days

Bavaria, 6.2.1978 (Pf)

Juv., Bavaria, June 1979 (Pf)

Bavaria, June 1979 (Pf)

) Bavaria, 1972 (Li)

Sand Martin (Riparia riparia)

The Sand Martin is the smallest European hirundine. It inhabits open country, particularly in the vicinity of water. The distribution is limited by the availability of suitable nest-sites.
The diet consists of small insects taken on the wing, often over water. The species is markedly gregarious, nesting in colonies and feeding and roosting in flocks. Outside the breeding season very large roosts are formed in reed-beds, willows and other marshy vegetation.
The nest-site is a tunnel in a sandy, gravel or clay bank. Both sexes excavate the tunnel which ends in a nest-chamber lined with feathers, grass stems and other plant material. Sometimes existing natural holes are used. In the north of the range the nest-sites may be very close to the ground, the banks of ditches and even low muddy or sandy sea-shores.
Incubation and care of the young are shared by both parents. Sand Martins are entirely migratory, wintering in Africa south of the Sahara.
Migration: Departs breeding areas in August–October, returning in March–May.

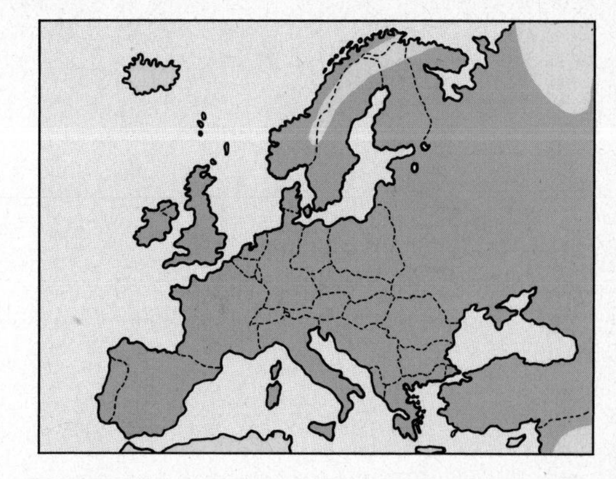

Length:	12 cm
Wing length:	10.1–11.0 cm
Weight:	14 g
Voice:	'Tchrrip', and a warning 'brrrit'. Twittering song
Breeding period:	Mid-May, June. 1–2 broods per year
Size of clutch:	4–5 (3–7) eggs
Colour of eggs:	White
Size of eggs:	18.0–12.6 mm
Incubation:	*ca* 14 days, beg. from complete clutch
Fledging period:	Nidicolous; young fly out at 16–20 days, return for another 14 days or so to the nest-hole

in clay bank, Bavaria, 1962 (Li)

Ad., feeding young, Roskilde, Denmark, 10.7.1964 (Chr)

Bavaria, nest exposed by landslip, 1979 (Li)

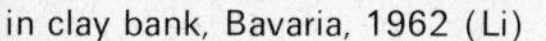

Pair at nest in turf wall, Sweden, (Sw)

Crag Martin (Ptyonoprogne rupestris)

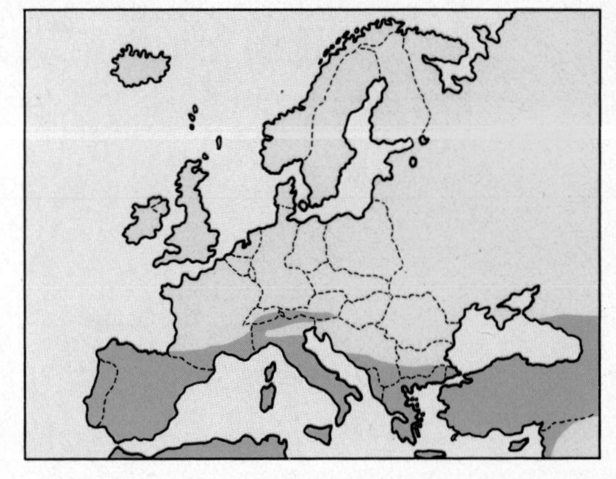

The Crag Martin occurs in rocky habitats, ranging from cliffs and river gorges in lowland areas to ravines and warm dry cliffs up to the snow-line in mountain areas. It generally requires a sunny and dry climate, but in recent years has extended its range to the north as far as the Channel coast.
It feeds over water and along the face of sunny rock-walls on small insects caught on the wing.
It is not particularly gregarious, breeding in loose colonies or singly. In winter it will form flocks at sites where food is abundant.
The nest is a half-cup made of mud and stuck to a vertical rock-face under a cornice, overhanging rocks, cave roofs, etc. The nest is sometimes lined with down, feathers and some plant material. Both parents take part in nest-building and rearing the young but the ♀ incubates alone.
The Crag Martin is resident and sedentary in some areas, probably where the winter microclimate is favourable. In other areas the species is migratory, wintering in parts of north and north-east Africa. The extent of the migratory movements is not fully understood. Local birds have been recorded as wintering as far north as the Alps, whilst it is absent from some areas of the Mediterranean in winter.
Migration: Mainly September–October, returning from late January to March.

Length:	15 cm
Wing length:	12.8–13.4 cm
Weight:	20–22 cm
Voice:	'Chitch' and 'tchrri'
Breeding period:	Mid-May to July. 1–2 broods per year
Size of clutch:	4–5 eggs, sometimes fewer
Colour of eggs:	Whitish, with reddish-brown to greyish spots and patches
Size of eggs:	21×14 mm
Incubation:	14 days, beg. from complete clutch
Fledging period:	Nidicolous; leaving the nest at 25 days and independent after a further 2–3 weeks

ta Valley, Carinthia, West Germany, 6.7.1977 (He)

Macedonia, Yugoslavia, 1969 (Li)

) Ad., Yugoslavia, 1971 (Li)

Swallow (Hirundo rustica)

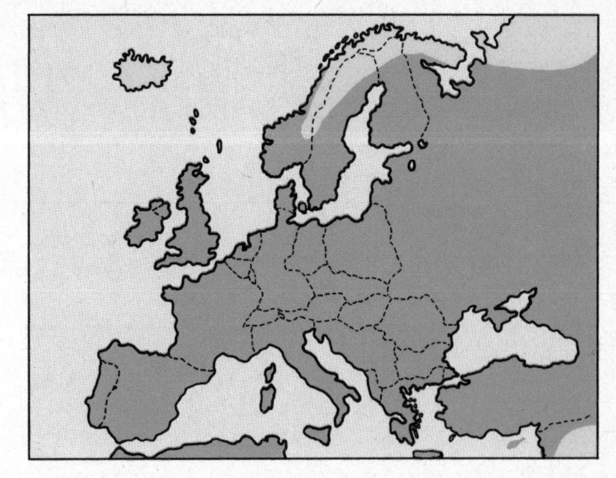

The Swallow is a familiar bird of open country, usually near farms, in villages and towns. It prefers to breed near water and dislikes densely forested areas. It feeds entirely on insects collected on the wing, often over water or near animals, where insects are plentiful.

Swallows are not very gregarious during the breeding season, though numbers are often seen together where insects are abundant. On migration and in winter they often form large flocks. Vast numbers have been recorded at reed-bed roosts.

The nest is built of mud mixed with grass, straw and other plant material. The birds collect wet mud from the banks of pools, lakes, etc. A lining of dry grass and feathers is provided. The nest is sited on a rafter in a shed or barn, under bridges, in buildings and, exceptionally, on the branch of a tree. Incubation is by the ♀ alone, with both parents rearing the young. The species is not colonial and generally only one nest is found at a particular site.

Swallows are migratory, wintering in Africa south of the Equator. A few birds occasionally winter as far north as Britain and more regular overwintering takes place in Mediterranean countries.

Migration: Departs breeding areas from July to October, returning in March–May.

Length:	19 cm
Wing length:	10–11.7 cm
Weight:	20 g
Voice:	'Tswit' and 'tswee'; chattering song of twittering and chirping noises
Breeding period:	End of May, June and July, beginning of August. 2–3 broods per year
Size of clutch:	5 (4–7) eggs
Colour of eggs:	White, with light or reddish brown spots
Size of eggs:	19.5 × 13.5 mm
Incubation:	14–16 days, beg. from complete clutch
Fledging period:	Nidicolous; fledged at 3 weeks, returning to nest to roost at first

aria, 1971 (Li)

Juv., Bavaria, August 1978 (Pf)

Bavaria, July 1978 (Pf)

♂ (*left*), ♀ (*right*), Bavaria, May 1979 (Pf)

Red-rumped Swallow

(Hirundo daurica)

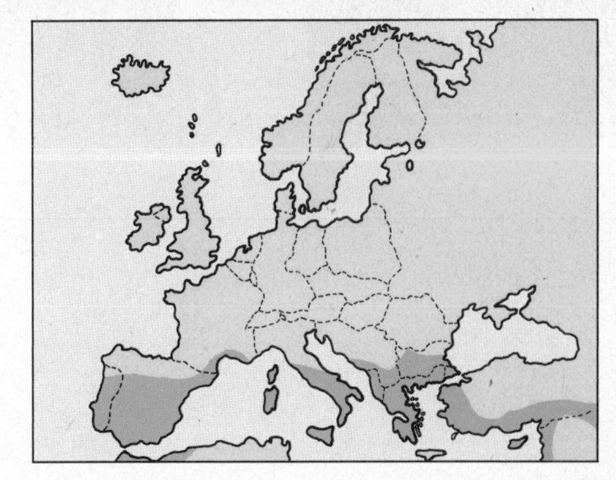

Red-rumped Swallows are found in dry sunny areas of limestone rock, mountains, sea-coasts and other areas where suitable nest-sites exist. Unlike the Swallow, they are rarely found near human habitation in the European range, though they occur in towns in the Middle East and North Africa.
The diet is entirely insects, caught on the wing and probably differs very little from that of the Swallow.
The nest-site is usually in a cave, rocky outcrop, under a bridge, or on a cliff, nearly always in the vicinity of water.
The nest is a round bowl with a tunnel entrance, built of mud pellets and plant material, lined with dry grass and feathers. It is fixed to the underside of a horizontal surface (cave-roof, etc.) Both sexes build the nest and care for the young, with the ♀ taking the major role in incubation. Nests are either single or in groups of four to five in a site.
The Red-rumped Swallow is entirely migratory, wintering in a belt across tropical Africa at about 10° N. It appears to be gradually extending its range in parts of south-west Europe.
Migration: Departs breeding areas in September–October, returning in late March–May.

Length:	19 cm
Wing length:	11.8–12.6 cm
Weight:	20 g
Voice:	Flight-call 'quitsch'. Song resembles Swallow but slower and less musical
Breeding period:	End of April to end of July. 2 broods per year
Size of clutch:	5 (3–7) eggs
Colour of eggs:	White
Size of eggs:	20 × 14 mm
Incubation:	14–16 days, beg. from complete clutch
Fledging period:	Nidicolous; fledged at about 3 weeks, returns to nest to roost for short periods

ece, 10.6.1979 (Li)

Nest under bridge, Greece, May 1975 (Fe)

Nest opened up by dogs, Greece, 11.6.1976 (Li)

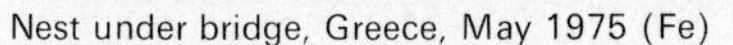

Ad. ♂ collecting mud for nest, Greece, May 1975 (Fe)

House Martin (Delichon urbica)

The House Martin shares the same open habitat as the Swallow but occurs in rocky and mountainous areas at a much higher altitude than that species. It also nests freely in towns and cities. It feeds exclusively on insects which are caught on the wing. It often associates with other hirundines when feeding but is inclined to feed at a higher altitude. It often perches on wires, buildings and bare trees.
There are two main types of nest-site. The natural one, where nests are fixed to cliff-walls, is commonest in the south of the range. In the north the artificial habitat of buildings is the most frequent, the nests being fixed just under the eaves. The species often nests in small colonies and, at times, the nests touch each other. The nest, built of mud with some plant material, is a rounded half-cup shape and lined with soft grass and feathers. Both sexes take part in nest-building, incubation and rearing the young. House Martins can be induced to colonise a building by the provision of artificial nests.
The species is entirely migratory, wintering in Africa. Its winter quarters are not well known, probably due to its habit of feeding at a high altitude above forested mountains.
Migration: Departs breeding areas in September–October, returning from March to June.

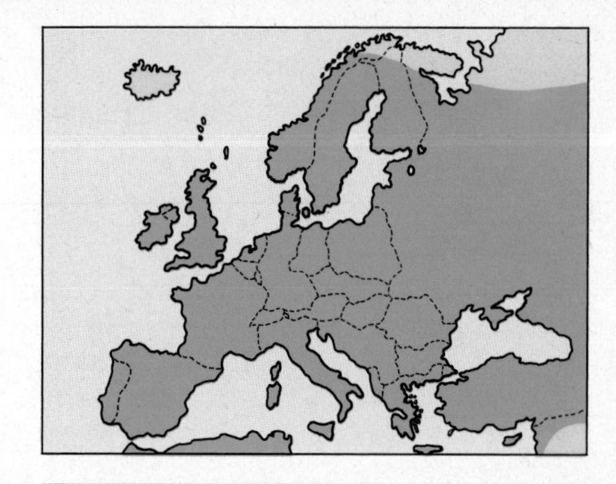

Length:	13 cm
Wing length:	10.4–11.4 cm
Weight:	20 g
Voice:	'Chirrpp'. Warning call: 'tseep'. Twittering song
Breeding period:	End of May, June, July, August. 1–2 broods per year
Size of clutch:	4–5 (2–6) eggs
Colour of eggs:	White
Size of eggs:	19×13 mm
Incubation:	13–19 days, beg. from complete clutch
Fledging period:	Nidicolous; flying at 19–25 days, but tended by parents for further 1–2 weeks

Denmark, June 1975 (Chr)

Collecting nest material, Bavaria, June 1979 (Pf)

Amrum Island, West Germany, 15.9.1972 (Qu)

♂ at half-finished nest, Sweden (Sw)

Blue-headed Wagtail

(Motacilla flava flava)

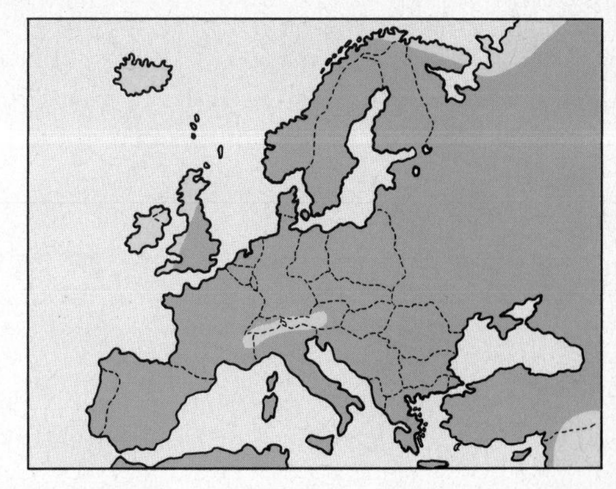

The Blue-headed Wagtail is the most widespread form of the Yellow Wagtail in Europe. It occurs throughout the middle of the Continent from southern Scandinavia to France and north Italy. It is a migrant in Britain and Ireland and has bred on occasion. In Britain it is replaced by the race termed Yellow Wagtail (*M. f. flavissima*).
The species is found mainly in lowland habitats; agricultural land, marshes, grassy meadows and similar areas, particularly near water, are frequented. The species feeds mainly on insects and small molluscs. The insects are taken on the wing as well as on the ground. Often Yellow Wagtails will associate with cattle to take advantage of the insects disturbed by the grazing animals. The nest is usually in a hollow in the ground in thick cover, sometimes in growing crops. The nest is a cup built of grasses, plant stems and roots with a lining of hair and a few feathers. Nest-building and incubation are done chiefly by the ♀, while both sexes share in rearing the young.
The species is entirely migratory, wintering in Africa (see overleaf). Outside the breeding season it is gregarious and very large numbers are met with at roosts, often in reed-beds and in the winter quarters.
Migration: Departs breeding areas in August and September, returning from March to May.

Length:	16.5 cm
Wing length:	8–8.5 cm
Weight:	17 g
Voice:	'Tsweep' or 'tsirr'; warbling song, widely varied by individuals
Breeding period:	May, June. 1 brood per year, 2 in the south of the range
Size of clutch:	4–6 (–7) eggs
Colour of eggs:	Whitish, with thick brown-grey speckling, appearing almost uniformly coloured
Size of eggs:	19×14 mm
Incubation:	12–13 days, beg. from complete clutch
Fledging period:	Nidicolous; remaining 11–12 days in nest, fledged at about 17 days

at nest, Amrum Island, 10.6.1969 (Qu)

Juv., Amrum Island, 10.6.1969 (Qu)

Nest lined with glass wool, Bremen, May 1968 (Sy)

ft) Ad. ♂, Bavaria, April 1973 (Pf)

Grey-headed Wagtail

(Motacilla flava thunbergi)

The Grey-headed Wagtail is the form which breeds in northern Scandinavia and parts of the Soviet Union. Its habits and breeding biology are much the same as for Blue-headed Wagtail, though the habitat includes swampy tundra, dward birch and willow scrub and areas with more trees than would be tolerated by the southern races.

Two other races of Yellow Wagtail which breed in Europe and are not illustrated are the Ashy-headed Wagtail (*M. f. cinereocapilla*), which breeds in Italy and parts of the Balkans, and the Spanish Wagtail (*M. f. iberiae*), which occurs in the Iberian peninsula.

There is much variation in all races of Yellow Wagtail and hybrid pairs often occur. The various races have separate winter quarters in Africa though certain races may share some areas. The Blue-headed Wagtail winters from Senegal and Nigeria to the Nile Valley and East Africa. The Yellow Wagtail winters mainly in Senegal and Gambia, the Grey-headed in a belt across northern tropical Africa and East Africa and the Spanish form in Morocco and coastal West Africa to Nigeria. The Ashy-headed race winters mainly in the Sudan and Chad.

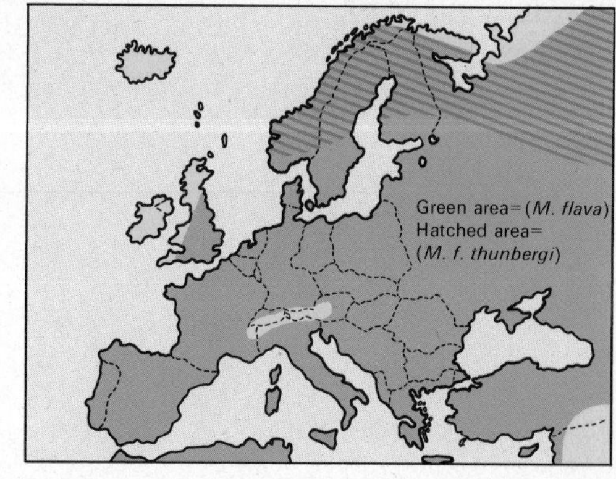

Length:	16–17 cm
Wing length:	8–8.5 cm
Weight:	17 g
Voice:	'Tseeh', and a twittering song with individual variations
Breeding period:	June. 1 brood per year. Replacement clutch possible
Size of clutch:	5–6 (–8) eggs
Colour of eggs:	Not easily distinguished from those of other races of Wagtail
Size of eggs:	18.5×14.1 mm
Incubation:	13–14 days, beginning from last egg
Fledging period:	Nidicolous; spending *ca* 11 days in the nest and being led a further few days

♂, Denmark (Ge)

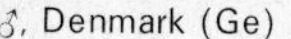

) Ad. ♀ at nest, Sweden (Sw)

Swedish Lapland, 23.6.1972 (Sy)

Black-headed Wagtail

(Motacilla flava feldegg)

This is one of the more distinctive races of Yellow Wagtail and has been given specific status by some authors in the past. Besides the more obvious plumage differences of the ♂, the ♀♀ tend to be nearly white underneath, often lacking the yellow tinge common to most other races. The call is louder and stronger than in other forms. The Black-headed Wagtail originated in Asia Minor and colonised parts of south-east Europe from there. It sometimes overshoots its breeding areas on spring migration and has occurred in Britain. It winters in Africa, mainly in the north-east of the continent and around Lake Chad. Like other Yellow Wagtails in winter, it is often found in association with herds of cattle, goats or big-game animals. They take insects disturbed by the hooves of the animals as well as directly from the backs of the animals themselves.

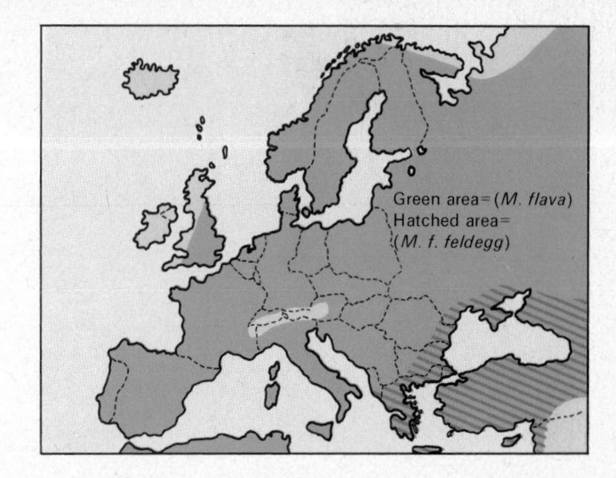

Length:	16.5 cm
Wing length:	7.9–8.8 cm
Weight:	20 g
Voice:	Loud and strident 'tswee', or 'tsweep'
Breeding period:	End of April, May, June. 1–2 broods per year. Replacement clutch possible
Size of clutch:	5–6 eggs
Colour of eggs:	Light sandy or grey, with thick or fine yellow-brown spots
Size of eggs:	18.7×14.4 mm
Incubation:	13–14 days, beg. from complete clutch
Fledging period:	Nidicolous; remaining 11–13 days in the nest, able to fly only at *ca* 17 days

♀, Greece, 25.5.1969 (Sy)

Nearly-fledged young, Greece, 22.6.1976 (Li)

) Ad. ♂, Greece, 25.5.1969 (Sy)

Grey Wagtail (Motacilla cinerea)

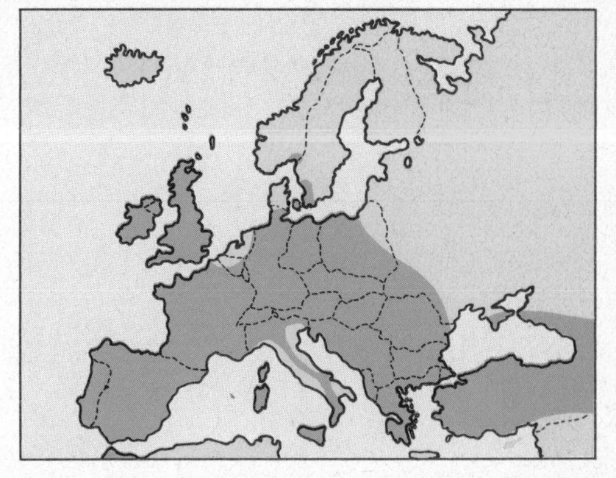

The Grey Wagtail is typically found on fast-flowing mountain streams. It will also occur on slower lowland waters but usually near a weir or waterfall. In mountainous areas it is found well above the tree-line and as high as 2,800 m. In winter it is more frequent on lowland waters such as watercress beds, sewage farms and the sea-coast.

It feeds mainly on insects taken in or near the water as well as some small crustaceans and even fish fry. It is generally seen singly or in pairs, and is not gregarious even on migration or in winter. The nest is usually sited amongst the roots of a waterside tree, on a ledge or in a hole in the bank. Sometimes a hole in a wall or an old Dipper's nest is used. The nest is a cup of grasses, twigs, roots, moss and other plant material, lined with hair, feathers or fine grass. The female takes the major part in nest-building and incubation but both sexes rear the young.

Grey Wagtails are mainly resident in western Europe, though there is some altitudinal movement in winter. Birds from the north and east of the European range migrate to winter along coasts and southwards to the Mediterranean. Some may go as far as Africa. Asian breeding birds are highly migratory, wintering in the tropics.

Migration: Mainly late September–November, returning in March–April.

Length:	18 cm
Wing length:	8–8.8 cm
Weight:	20 g
Voice:	Common call: 'tsississ'. Warning note: 'siz-eet'. Song: variable musical twittering
Breeding period:	April, May, June. 1–2 broods per year
Size of clutch:	4–6 (3–7) eggs
Colour of eggs:	Yellowish white, with dense red-brown spots
Size of eggs:	19×14 mm
Incubation:	12–14 days, beg. from complete clutch
Fledging period:	Nidicolous; leaving nest at 12–13 days, but unable to fly for a further 3–4 days

♂ at nest, Bavaria, 1972 (Li)

Juvs, Bavaria, 31.5.1977 (Li)

Bavaria, 1975 (Li)

) Ad. ♀, Bavaria, April 1972 (Pf)

White Wagtail (Motacilla alba alba)

There are two forms of the White Wagtail breeding in Europe; the nominate (*alba*), which breeds over most of continental Europe and Iceland, and the Pied Wagtail (*M. a. yarrellii*), which breeds in the British Isles and Ireland. Interbreeding has been recorded. The White Wagtail occurs in a wide variety of habitat. It likes grassy areas, parks, large gardens, agricultural land, areas near water and farmyards. It feeds mainly on insects and other small invertebrates.

The nest-site may be in a hole on a bank or cliff, in a wall, building, pipe, old nest of other birds, under bridges and on the ground. The nest is a cup built of grass, stems and other plant material and is lined with hair, feathers or wool. It is constructed entirely by the ♀, who also takes the main share of incubation. Outside the breeding season the species is gregarious and often roosts in large numbers. These roosts are sometimes in trees in city streets, in glasshouses or in reed-beds.

The White Wagtail is migratory in the north and east of its range, wintering from southern Europe to North Africa and as far south as the Equator in East Africa. The Pied Wagtail is mainly resident.

Migration: Departs breeding grounds in August–October, returning in March–May.

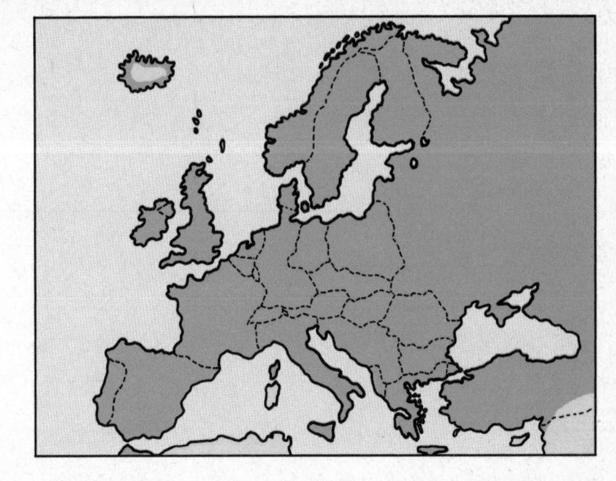

Length:	18 cm
Wing length:	8.3–9.3 cm
Weight:	23 g
Voice:	'Tchizzik'. Song: a warbling twitter
Breeding period:	Beginning of May, June, mid-July. 2 broods per year
Size of clutch:	5–6 (4–7) eggs
Colour of eggs:	Whitish, with dark grey and brown spots
Size of eggs:	20.5 × 15.3 mm
Incubation:	12–14 days, beg. from complete clutch
Fledging period:	Nidicolous; leaving the nest at 14–15 days, and tended a further 15 days or so by their parents

., Bavaria, August 1973 (Pf)

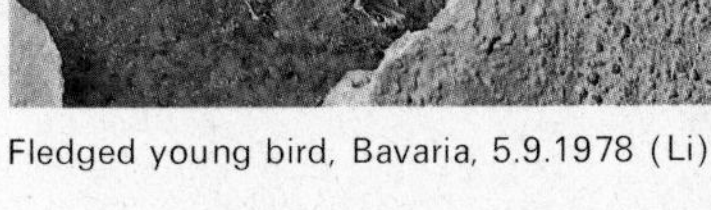

Fledged young bird, Bavaria, 5.9.1978 (Li)

Bavaria, 1973 (Li)

) *M. a. alba*, Bavaria, 1975 (Li)

Tawny Pipit (Anthus campestris)

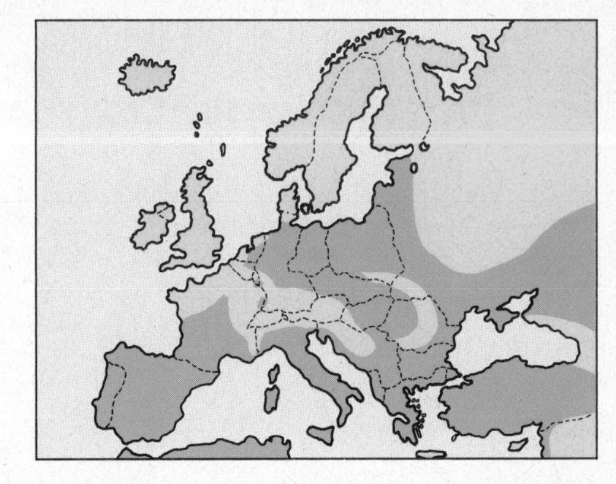

The Tawny Pipit frequents dry open areas with sparse vegetation such as sandy heaths, river banks, mountain slopes, fallow cultivated land, grassy areas with light soils and vinyards.
It feeds mainly on insects and larvae, spiders and other small invertebrates. Outside the breeding season it is quite gregarious and occurs in small parties, especially when associating with grazing cattle (cf. Yellow Wagtail). The ♂ has a high, circling display-flight ending in a plummet to the ground. It sings both in flight and from prominent perches such as wires or bushes. The nest-site is in a hollow on the ground, usually in the cover of some low vegetation. The nest is a cup of grass and other plant materials lined with fine grass or hair. Usually the ♀ is responsible for nest-building and incubation but the latter, at least, is shared in some pairs. Both sexes feed the young.
The Tawny Pipit is entirely migratory, wintering in dry desert-like areas of Africa in a belt south of the Sahara and in north-east Africa.
Migration: Departs the breeding areas in late August–October, returning in April–May.

Length:	17 cm
Wing length:	8.1–9.8 cm
Weight:	25 g
Voice:	'Chirrup' and 'tzic' or 'sweep'. Song: 'chiree' repeated
Breeding period:	Mid-May to July. 2 broods per year
Size of clutch:	4–5 (3–6) eggs
Colour of eggs:	Whitish with a dense smattering of large or small, middle- to light-brown spots and patches
Size of eggs:	21.5×16 mm
Incubation:	13–14 days, beg. from complete clutch
Fledging period:	Nidicolous; leaving the nest at 12–14 days, still unable to fly for a few days

‹ey, 28.6.1977 (Li)

Fledged young bird, Turkey, 1977 (Schu)

Turkey, 21.6.1977 (Schu)

) Turkey, 29.6.1977 (Li)

Tree Pipit (Anthus trivialis)

The Tree Pipit is found in areas of dry grassland with scattered trees, heathland, open woodland or woodland edge, especially areas where trees have been burnt or cleared.
It feeds mainly on insects and larvae, spiders and other invertebrates; some seeds are taken at times.
The species is gregarious outside the breeding season and large flocks occur on migration and in winter.
The ♂ has a special song-flight. Taking off from a tree, it flutters steeply upwards and parachutes down with wings raised and tail spread.
The nest-site is a hollow in the ground, usually in the cover of low vegetation. The nest is a cup built of grass, moss and other plant material. The construction of the nest and incubation are left to the ♀ alone while both sexes share with rearing the young.
The Tree Pipit is entirely migratory, wintering in tropical Africa.
Migration: Departs breeding areas in August–October, returning in April–June.

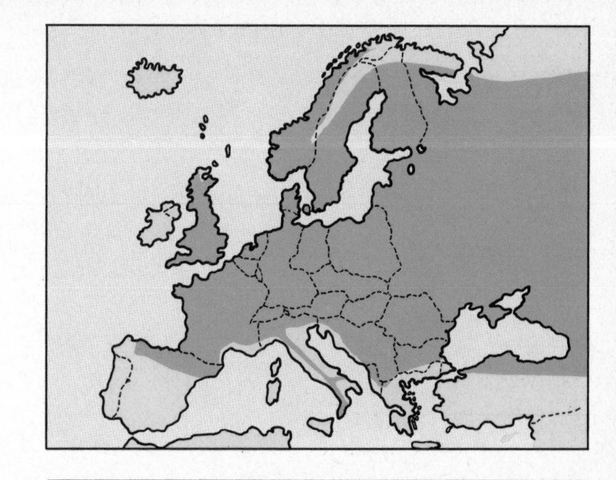

Length:	15 cm
Wing length:	8.0–9.0 cm
Weight:	22 g
Voice:	'Tseez', and alarm 'sip, sip, sip'. Song: loud and musical canary-like 'tseeatseeatseea'
Breeding period:	Early May to early July. 1–2 broods per year
Size of clutch:	5 (4–6) eggs
Colour of eggs:	Very variable. Whitish to red-brown, with dense spots of brown, olive, black, etc.
Size of eggs:	20.4×15.4 mm
Incubation:	12–14 days, beg. from complete clutch
Fledging period:	Nidicolous; leaving the nest at *ca* 12 days, not able to fly for further 2–3 days

song-flight, Bavaria, June 1977 (Wo)

Fledged young bird, Bavaria, 19.6.1979 (Li)

Württemberg, West Germany 1976 (Ca)

t) Austria (He)

Meadow Pipit (Anthus pratensis)

The Meadow Pipit is a bird of open country. Moorland, heath, marsh, dunes and rough pasture are all typical habitats. It is also found in mountains, even above the tree-line, and in the Arctic, where it inhabits shrub tundra. In winter it is common on coastal grassland and agricultural land. The diet consists of small insects, spiders, earthworms and some seeds. Outside the breeding season the species is gregarious and large numbers are often seen together, particularly on migration. The ♂ has a rather feeble song-flight, rising from the ground and gliding down again like a Tree Pipit. The nest-site is on the ground, usually well concealed by vegetation. The nest is a cup of dry grasses and other plant matter, lined with hair or fine grass. Incubation is by the ♀ alone, though both sexes rear the young.
The Meadow Pipit is migratory, at least in the north and east of the range, and winters in western Europe and Mediterranean countries. In some areas it is more or less resident but leaves high ground for coastal areas in winter.
Migration: August–November, returning in February–May.

The **Red-throated Pipit** (*Anthus cervinus*) is the northern counterpart of the Meadow Pipit. It is found in marshy tundra and willow or birch scrub. In Scandinavia it nests alongside Meadow Pipit in the same habitat. It is entirely migratory, wintering in Africa.

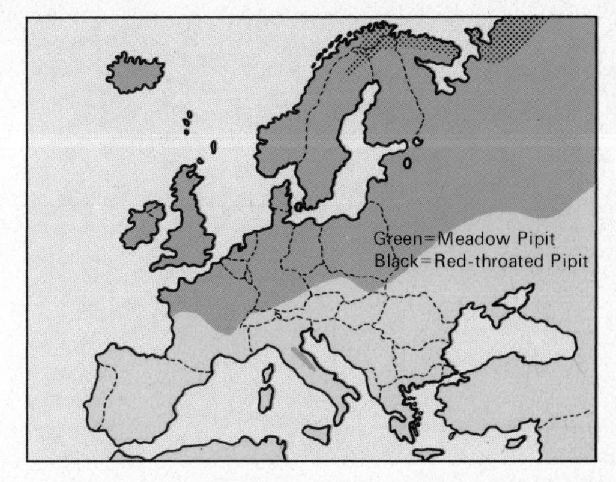

Length:	15 cm
Wing length:	8–8.4 cm
Weight:	20 g
Voice:	Call: 'tsiip, tsiip'. Song a tinkling trill: 'tsitsitsitsiss'
Breeding period:	End of April to end of July. 2 broods per year
Size of clutch:	4–5 (6) eggs
Colour of eggs:	Whitish to light green, densely flecked with shades of brown, most dense at the blunt pole
Size of eggs:	19.5×14.4 mm
Incubation:	14 days, beg. from complete clutch
Fledging period:	Nidicolous; leaving the nest at 12–14 days still unable to fly

ıdow Pipit ♀ at nest, Amrum Island, 20.5.1974 (Qu)

Meadow Pipit's clutch, Sweden, June 1978 (Ca)

Red-throated Pipit, 1.4.1971 (Haas)

) Meadow Pipit, Sweden, June 1978 (Ca)

Water Pipit (Anthus spinoletta)

There are two distinct groups within this species, the Water Pipit (*A. s. spinoletta*) and the Rock Pipit (*A. s. petrosus/littoralis*, etc.). The former is found in the mountains of central and south Europe up to a height of 8,000 feet. It breeds in alpine meadows and rocky scrub above the tree-line. It feeds mainly on small insects.
The nest is in a hollow in the ground, often on a steep slope and in the cover of low vegetation. Nest-building and incubation are the responsibility of the ♀. Both sexes rear the young. The nest is constructed of grass, plant stems and other plant materials, lined with fine dry grass or hair. Water Pipits leave the higher altitudes in winter and either remain in southern Europe or migrate north of the breeding range to winter in Britain and north continental Europe, where they occur at sewage farms, gravel pits, marshes, etc. The **Rock Pipits** are a coastal form, breeding in north-west Europe. They frequent rocky sea-shores and islands, often nesting in rock-crevices, under driftwood or on 'inland' sites in heath or moorland. The diet includes small molluscs, crustaceans and littoral débris. The Scandinavian race is migratory and winters along the North Sea and Channel coasts. It is found on salt-marshes and coastal pastures as well as on the tideline. The British race is largely resident.
Migration: Departs breeding areas in August–October, returning in March–May.

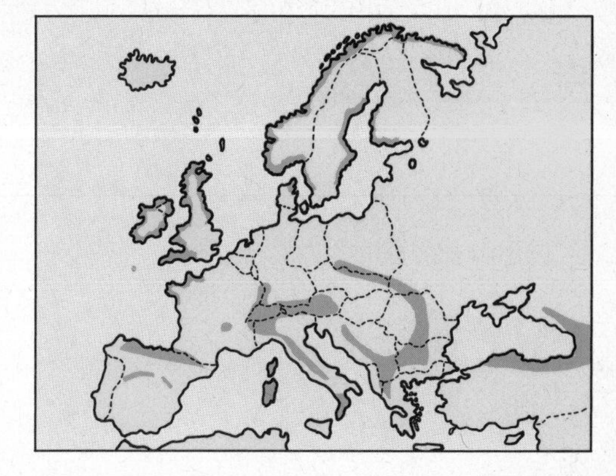

Length:	17 cm
Wing length:	8.0–9.6 cm
Weight:	19–28 g
Voice:	'Tsip-eep'. Alarm: 'Jeet'. Song: 'Tleetleetleetlee'
Breeding period:	End of April to July. 1 brood per year in the north, 2 in the south of Europe
Size of clutch:	4–5 (–6) eggs
Colour of eggs:	Whitish, strongly speckled with various shades of brown
Size of eggs:	21.3×15.5 mm
Incubation:	14 days, beg. from complete clutch
Fledging period:	Nidicolous; remaining 14 days in nest and flying in a further 2 days

'ged young bird, Austria, 23.7.1978 (Li)

Young in nest, Austria, 23.7.1978 (Li)

Clutch with Cuckoo's eggs (brown), Bavaria, 25.5.1976 (Li)

) *A. s. spinoletta*, Austria, 29.7.1978 (Li)

Red-backed Shrike (Lanius collurio)

The Red-backed Shrike inhabits heathland with scrub and scattered trees, hedgerows, light woodland or forest edge in lowland and hilly regions to about 1,800 m.

The diet includes a variety of insects, especially beetles and grasshoppers, small birds mostly up to the size of House Sparrow, chicks of other birds including Pheasant and partridges, mice, shrews, amphibians and small lizards. The prey is usually obtained by perching in a prominent place (wires, posts or bushes) and swooping down to take food from the ground. Small birds may be taken on the wing and an aerial chase may take place. The Red-backed Shrike often forms so-called 'larders', where prey items are impaled on thorns or barbed wire. Sometimes the surplus food is left uneaten. The nest-site is usually in a thick bush, often a thorn bush, and is about 1–3 m from the ground. The nest is a large cup of grass, plant stems and twigs and the lining is fine grass, hair and wool or down. The ♂ builds the main structure and the ♀ finishes the lining. The ♀ broods alone, being fed by the ♂, while both sexes rear the young.

The Red-backed Shrike is entirely migratory, wintering in East and South Africa.

Migration: Departs breeding areas in August–September, returning in late April–June.

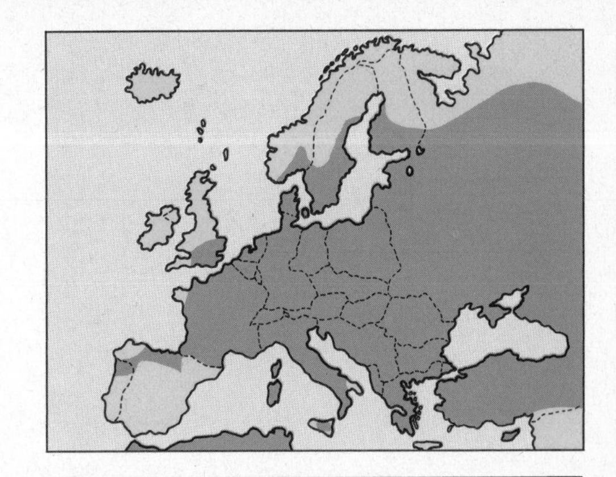

Length:	17 cm
Wing length:	8.9–9.8 cm
Weight:	30 g
Voice:	'Chee-uk' or 'shak'. The soft warbling song includes imitations of other birds
Breeding period:	Mid-May to June. 1 brood per year. Replacement clutch possible
Size of clutch:	5–6 (4–7) eggs
Colour of eggs:	Very variable. Greenish, yellowish and reddish. A crown of thick spots of various browns at the blunt pole
Size of eggs:	23×17.1 cm
Incubation:	15 days, beg. from penult. or last egg
Fledging period:	Nidicolous; leaving the nest at 15 days but tended by both parents for a further 3–4 weeks

♀ at nest, Bavaria, 1960 (Li)

Bavaria, June 1973 (Pf)

Turkey, 31.5.1973 (Li)

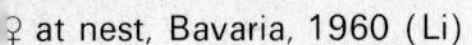

♂ and ♀ at nest, Bavaria, June 1972 (Pf)

Lesser Grey Shrike (Lanius minor)

The Lesser Grey Shrike frequents open country with scattered trees and bushes, It also occurs in large gardens, roadside areas, hedgerows and agricultural land. It prefers hot, dry conditions and does not occur in high mountainous areas.
The diet is mainly large insects, particularly beetles, and it will take some small mammals as well as fruit such as cherries and figs. It spots its prey from an exposed perch (frequently wires) and glides down to pounce on its prey. It will impale the prey on thorns but the habit of forming larders is not as frequent as in the Red-backed Shrike.
The nest-site is usually in a tree, often up to 20 m from the ground. Where large trees are absent it will nest in bushes. The nest is usually placed next to the trunk or a large branch and is a substantial cup of twigs, grass and other plant material, lined with fine grass, hair, wool or feathers. The ♀ does most of the nest-building and takes the larger share of incubation. Both sexes rear the young. The species is aggressive in defence of the nest-site and will attempt to drive off other birds even as large as the Buzzard.
The Lesser Grey Shrike is a migrant, wintering in southern Africa.
Migration: Autumn dispersal from September to October, returning in late April–June.

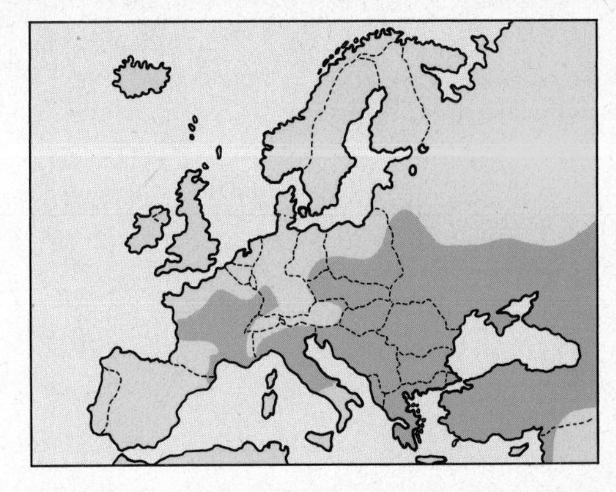

Length:	20 cm
Wing length:	11.3–12.3 cm
Weight:	45–60 g
Voice:	'Shek-shek'. Chattering song, and imitates the voices of other birds
Breeding period:	End of May, June. 1 brood per year. Replacement clutch possible
Size of clutch:	5–7 eggs
Colour of eggs:	Light greenish, with sparse heavy patches of light or olive-brown and grey, more at the blunt pole
Size of eggs:	24.5×18 mm
Incubation:	15 days, beg. from complete clutch
Fledging period:	Nidicolous; leaving the nest, scarcely able to fly, at 14 days, and tended for a further period

♂, Neusiedler See, Austria, 12.5.1964 (Sy)

Imm., Austria, 1968 (Li)

Macedonia, Yugoslavia, 1969 (Li)

) Austria, 1968 (Li)

Masked Shrike (Lanius nubicus)

The Masked Shrike has a very limited distribution in south-east Europe and Asia Minor. It inhabits dry areas with olive groves, scattered bushes and hedges. It also occurs in light woodland with open clearings. In some areas it is found alongside Woodchat Shrikes, even nesting as close as 15 m from the nest of that species. It feeds almost entirely on insects, particularly ants and beetles. It takes flying insects in a flycatcher-like manner, darting and swooping from a tree or bush.

The nest-site is in a tree or large bush, sometimes as high as 10 m from the ground. The nest is constructed from roots, grass, plant stems, etc. and lined with feathers and fine plant-fibres. It is often placed in the fork of a leafless branch or close to the trunk of a tree. Both sexes share in incubation and feeding the young, with the ♂ providing most of the food.

The species is rarely seen away from cover and is more secretive than the other European shrikes, perching less frequently in prominent exposed sites. The Masked Shrike is a migrant, wintering in a small area of north-east Africa, mainly the Sudan and Somaliland.

Migration: Departs breeding areas in September–October, returning in April and May.

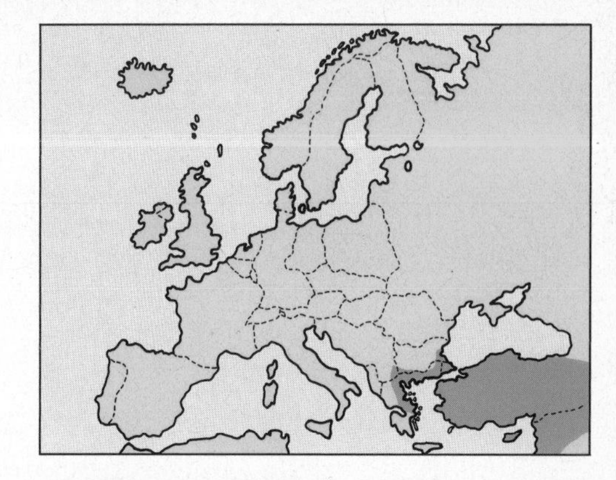

Length:	17 cm
Wing length:	8.7–9.2 cm
Weight:	35 g
Voice:	'Keer'. Song: soft scratching notes
Breeding period:	End of May, June. 1 brood per year. Replacement clutch possible
Size of clutch:	4–7 (3) eggs
Colour of eggs:	Creamy yellow to yellowish brown, with light and dark brown spots in a zone round the blunt pole
Size of eggs:	20.8×16 mm
Incubation:	15 days, beg. from complete clutch
Fledging period:	Nidicolous; leaving the nest not quite able to fly. Both parents lead the brood

eft), ♂ (*right*), Greece, 2.7.1977 (Li)

Almost-fledged young, Greece, 1.7.1977 (Li)

Greece, 3.5.1978 (Li)

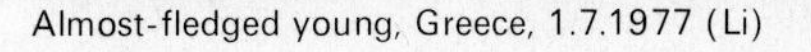

Ad. ♂ at nest, Greece, 2.7.1977 (Li)

Woodchat Shrike (Lanius senator)

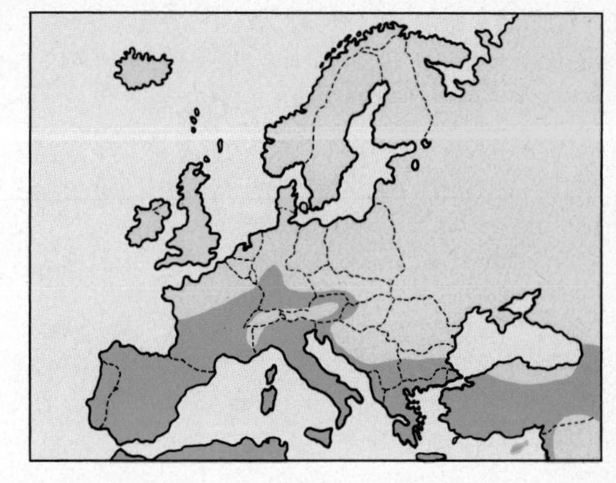

The Woodchat Shrike inhabits open woodland, parkland, gardens, olive groves and scrubby areas. It is found in both lowland and mountainous areas and overlaps with both Red-backed and Masked Shrikes in choice of habitat. The diet is mainly large insects including bees, grasshoppers and beetles. Small birds and their young are also taken. It spots its prey from a high observation point, often in a tree, though wires and posts are also used.
The nest is placed in a tree or shrub up to 4 m from the ground. It is built of roots and other plant material lined with feathers, hair and wool. It is often a substantial structure. Both sexes take part in nest-building but the ♀ incubates alone. The ♂ provides food for her and both sexes feed the young. Where food is abundant the species may nest quite close together, sometimes only 100 m apart.
The ♂ has the best song of all the European shrikes, usually singing from the cover of a bush or tree.
Woodchat Shrikes are entirely migratory, wintering in belts across tropical Africa.
Migration: Departs European breeding areas as early as July, but mainly August–September, returning in April–May.

Length:	17 cm
Wing length:	11.3–11.6 cm
Weight:	40 g
Voice:	'Kiwick'. Song: rich warbling, contains striking imitations of other birds
Breeding period:	End of April to June. 1 brood per year. Replacement clutch possible
Size of clutch:	5–6 (4–8) eggs
Colour of eggs:	Pale green, with spots in various shades of brown in zone at the blunt pole
Size of eggs:	23×17 mm
Incubation:	16 days, beg. from complete clutch
Fledging period:	Nidicolous; leaving the nest at the age of *ca* 16–19 days

t nest, 30.5.1978 (Li)

Fledged young bird, Württemberg, 1977 (Schu)

Macedonia, Yugoslavia, June 1971 (Li)

t) ♂ feeds brooding ♀, Greece, 3.5.1978 (Li)

Great Grey Shrike (Lanius excubitor)

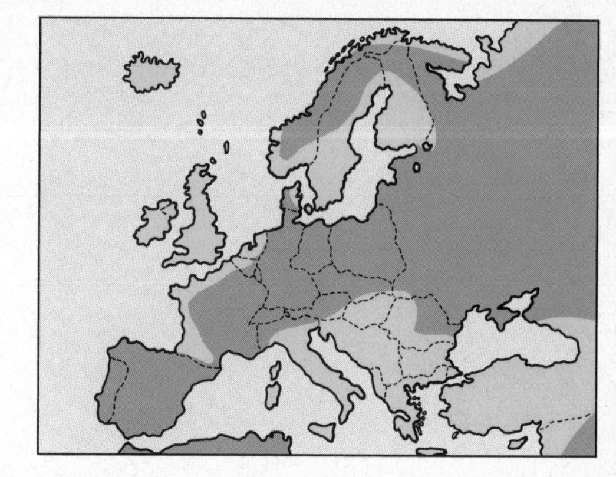

Because of its wide distribution in Europe the Great Grey Shrike is found in a variety of different habitats. In the north it inhabits shrub tundra, birch woods, heath and moorland with scattered trees, open forest and forest clearings. In the south it occurs in dry scrub, orchards, gardens, semi-deserts and woodland. It feeds mainly on small birds, mostly passerines, lizards, small mammals, large insects and some amphibians. It spots its prey from an exposed perch, often wires or posts, and will give chase to small birds in flight. It also takes prey from the ground and frequently hovers when looking for insects or small mammals. Its prey is often impaled on thorns or jammed in the fork of a branch, larders being formed in both summer and winter quarters. The nest is a large cup of twigs, plant stems, grass and moss, lined with fine plant material, hair or feathers. Both sexes collect material but only the ♀ builds. Incubation is mainly by the ♀ with both sexes rearing the young. The species is aggressive in defence of the territory, frequently driving off much larger birds.

The Great Grey Shrike is entirely migratory in the north of its range, wintering in Britain and throughout the southern breeding areas, North Africa and the Middle East.

Migration: Departs northern areas in September–November, returning in April–May.

Length:	24 cm
Wing length:	10.7–11.8 cm
Weight:	60–70 g
Voice:	'Shek-shek', 'jaa-eg'. Song: mixture of harsh and musical notes; imitative
Breeding period:	April, May, June. 1 brood per year. Replacement clutch possible
Size of clutch:	5–7 (–9) eggs
Colour of eggs:	Variable; bluish to greenish white, densely patched with large blobs in range of browns
Size of eggs:	26.3×19.3 mm
Incubation:	15 days
Fledging period:	Nidicolous; flying at 19–20 days, independent after a further 15 days

aria, 1971 (Li)

Fledged young bird, Bavaria, June 1959 (Pf)

Bavaria, 1971 (Li)

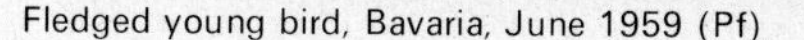

') ♂ feeds brooding ♀, Bavaria, 1971 (Li)

Waxwing (Bombycilla garrulus)

The Waxwing is one of the characteristic birds of the northern taiga forests. It breeds in dense coniferous forest of spruce, pine and firs with an understorey of berry-bearing bushes. It also nests in scattered trees along the edge of moorland or in river valleys. In winter it is found in any area with berry-bearing plants, open country, gardens, hedgerows, etc.
The diet includes small insects, caught on the wing in the manner of a flycatcher and a wide variety of berries such as bilberry, crowberry, juniper, etc. In winter, berries are the main food. Outside the breeding season it is often gregarious and may occur in large flocks where food is plentiful.
The nest is usually in a conifer, up to 6 m from the ground. It is built of conifer twigs with moss, lichens and feathers. It is built mainly by the ♀ who also takes the main share of incubation, being fed by the ♂. Both sexes feed the young on a diet of regurgitated insects and berries.
Waxwings are mainly migratory, leaving the northern forests in winter. When food is scarce considerable movements or 'invasions' take place, with birds occurring far to the south or west of the breeding areas. In other years the movements are less spectacular, with birds wintering in north and central Europe.

Migration: Departs northern forests in October onwards, movements taking place throughout the winter as food supplies become exhausted. Spring passage in March–April.

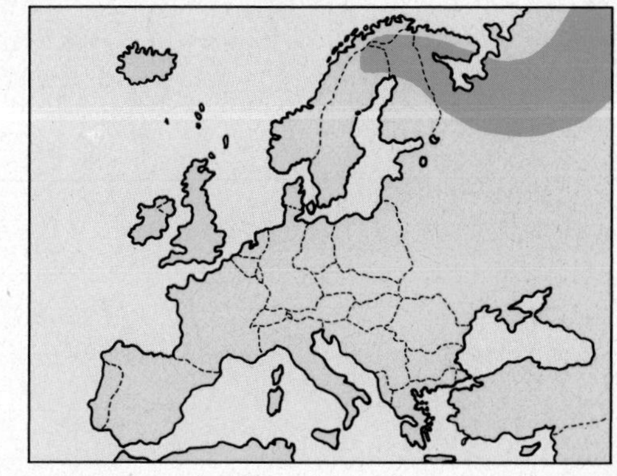

Length:	18 cm
Wing length:	11.2–12 cm
Weight:	60 g
Voice:	'Zhreee' or 'sirrr'. Song: a high trill
Breeding period:	June. 1 brood per year
Size of clutch:	5 (4–6) eggs
Colour of eggs:	Light grey to grey-brown, with sparse black or grey spots
Size of eggs:	24.6×17 mm
Incubation:	14–17 days, beginning from last egg
Fledging period:	Nidicolous; fledged at *ca* 16 days

er Saxony, 14.12.1963 (Pl)

Fledged young bird, Finland (Hau)

) Ad. with juv., Finland (Hau)

Dipper (Cinclus cinclus)

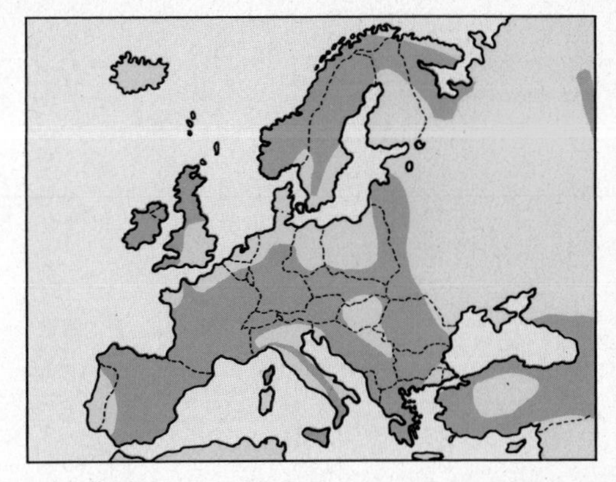

The Dipper is found by mountain streams and rivers and sometimes by the shallow margins of lakes or tarns. It will occur in some lowland areas if there is swift-flowing water and a suitable nesting-site such as a bridge or mill. In some areas it occurs on the lower reaches of rivers and even the coast in winter. It feeds on the larvae of aquatic insects such as caddis fly, mayfly, etc. as well as on small crustaceans, worms, beetles, tadpoles and small fish. It obtains its food by wading, swimming or diving into the water and deliberately walking underwater in search of food items. It is often seen perched on a rock or stone protruding from the water. It has a characteristic bobbing movement, jerking its tail downwards.

The nest-site is in a hole in the bank, among rocks or masonry such as a bridge or culvert. Sometimes a cave behind a waterfall is used. The nest is a large domed structure made of moss, leaves and small twigs, fitting exactly into a hole or sometimes on a support such as a ledge or branch. Both sexes share in nest-building. Incubation is entirely by the ♀. She is rarely fed by the ♂ but leaves the nest for short periods to feed. Both parents feed the young. The species is highly territorial, even in winter, and is usually seen singly or in pairs.

It is mainly resident and sedentary throughout the range but some movement of Scandinavian birds takes place with dispersal to lower altitudes and the coast. Some Scandinavian birds have reached Britain to winter.

Length:	18 cm
Wing length:	8.5–9.6 cm
Weight:	60 g
Voice:	'Zit', 'clink', liquid warble with loud explosive notes
Breeding period:	Mid-March, April and June–July. 2 broods per year. Replacement clutch possible
Size of clutch:	5 (3–8) eggs
Colour of eggs:	White
Size of eggs:	26 × 18.5 mm
Incubation:	15–18 days, beg. from complete clutch
Fledging period:	Nidicolous; leaving nest at 19–25 days; young can swim before they can fly

aria, 1970 (Li)

Free-standing nest, Bavaria, 15.4.1970 (Li)

Juv., North Hessen, 21.5.1975 (Tö)

Ad., Odenwald, Hessen, 6.4.1963 (Pf)

Wren (Troglodytes troglodytes)

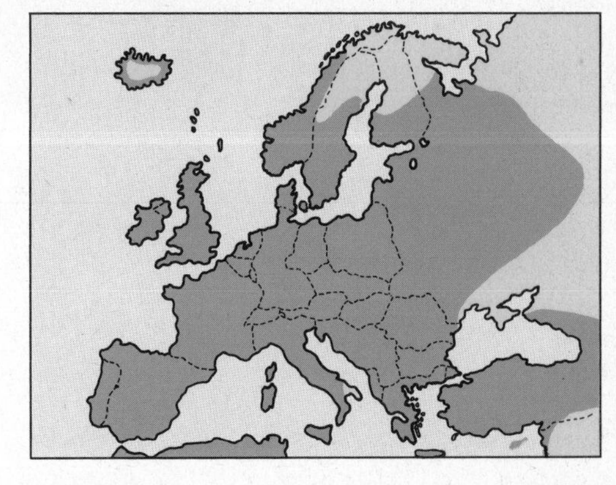

The Wren is a common bird of woodland areas with dense undergrowth, hedges, gardens, river banks and other areas with low cover. In some offshore islands in northern Europe it is also found on sea-cliffs and rocky areas. In winter it is common in reed-beds.
The Wren feeds mainly on small insects and larvae, spiders and, occasionally, seeds. The young are fed mainly on larvae, particularly of moths.
The nest-site is any type of hollow, crevice or hole in steep banks, trees, among rocks and sometimes in unlikely sites such as discarded cooking utensils (kettles, etc.). The nest is a beautifully constructed dome of moss, leaves, lichen and plant material lined with feathers. The ♂ builds the main structure and the ♀ lines the nest. The ♂♂ may build several nests within a territory and, being polygamous, may have several ♀♀ incubating in different nests. The nests are sometimes used for roosting outside the breeding season. The incubation is by the ♀ alone with both sexes feeding the young. Where the ♂♂ have several ♀♀ nesting within the territory the broods hatch at intervals so that the ♂ helps first with one brood then another in succession. Wrens are mainly resident and sedentary and are subject to severe losses in hard winters. Northern birds seem to migrate and there is a passage of Wrens at most coastal observatories. The extent of the movements is not known partly due to the Wren's secretive behaviour and the dense cover the species inhabits.
Migration: Coastal passage occurs mainly in September–October and in March–April.

Length:	10 cm
Wing length:	4.3–5.5 mm
Weight:	9 g
Voice:	'Tit tit tit', loud and strident; clear rattling warble
Breeding period:	End of April, May and June, July. 2 broods per year. Replacement clutch possible
Size of clutch:	5–8 (10) eggs
Colour of eggs:	White, with a few fine rusty to dark brown spots
Size of eggs:	17.6×13.3 mm
Incubation:	14–17 days, beg. from complete clutch
Fledging period:	Nidicolous; leaving nest at 15–20 days; led a short while longer

aria, 1965 (Li)

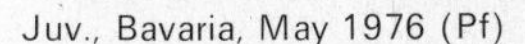

Juv., Bavaria, May 1976 (Pf)

Nest opened by Jay, Bavaria, 1973 (Li)

Ad. at nest, Bavaria, May 1976 (Pf)

Alpine Accentor (Prunella collaris)

The Alpine Accentor is found in the high mountains of central and southern Europe. It occurs between the tree- and snow-lines, frequenting alpine meadows, rocky slopes (sometimes with sparse scrub vegetation), bare rock-faces and scree. It prefers dry, sunny locations.
It feeds on a wide variety of insects and larvae, spiders, small molluscs and worms. In winter, seeds of grasses and other low plants are important. It can often be seen along the edge of the snow-line in spring taking food items from the melting snow.
It is a quiet and unobtrusive bird, hopping or walking among rocks and boulders. The ♂ has a short song-flight, similar to a lark's, but also sings from the top of a boulder or small bush.
The nest is usually in a crevice or among boulders though sometimes in the shelter of low vegetation. It is a neat cup of roots, plant stems and similar materials, lined with moss, lichen or fine grass. Both sexes take part in incubation and rearing the young. Outside the breeding season it is sometimes quite gregarious and small parties are frequently seen together. Though mainly resident, it usually descends to a lower altitude in winter, normally below the snow-line. Some movements do occur, with records from lowland areas well outside the breeding range, including Britain on rare occasions.

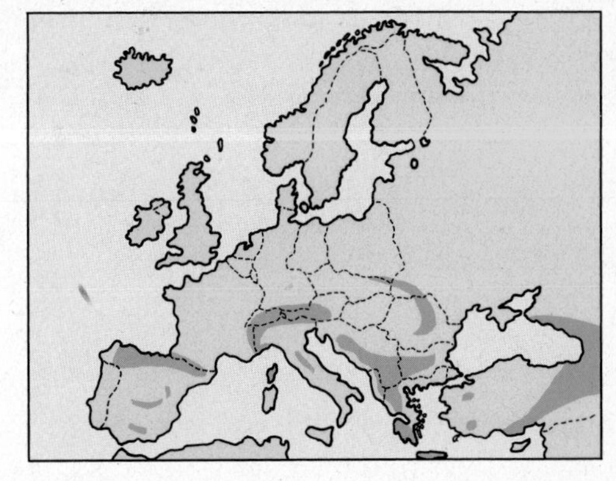

Length:	18 cm
Wing length:	10.3–10.8 cm
Weight:	40 g
Voice:	Call: 'tehirr-rie'. Song: like Woodlark, with trilling and piping
Breeding period:	Mid-May, June and July–August. Usually 2 broods per year
Size of clutch:	4–5 (3–7) eggs
Colour of eggs:	Turquoise
Size of eggs:	23×16 mm
Incubation:	15 days, beg. from complete clutch
Fledging period:	Nidicolous; leaving the nest at 16 days, before able to fly

ɘeding, Carinthia, 24.7.1978 (He)

Mating, Carinthia (Aich)

Carinthia (Aich)

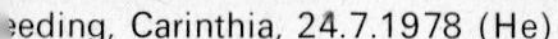

) Ad., Carinthia, West Germany (Aich)

Dunnock (Prunella modularis)

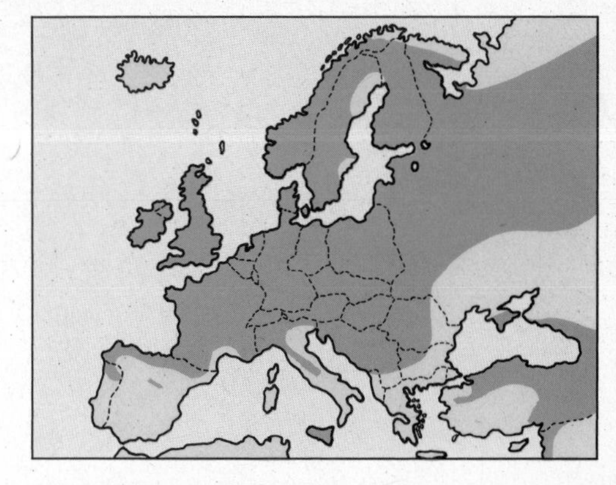

The Dunnock is a common bird in both coniferous and broad-leaved woodland, parks, gardens, hedgerows and bushy or scrub areas. It is an unobtrusive bird, feeding mainly on the ground in a characteristic manner. It feeds largely on seeds of plants such as chickweed, plantain, spurrey, etc. and on a variety of small insects, spiders and small worms. In winter seeds are the main food item.

The nest-site is usually well hidden in a tree, shrub or low cover and sometimes the old nests of other birds are used. The nest is a deep cup of twigs, plant stems, grasses and other plant material, lined with moss, hair or wool and sometimes a few feathers. The nest is constructed by the ♀ alone, and she also undertakes incubation alone, leaving the nest for short periods to feed. Both sexes share in rearing the young.

Throughout most of Europe the Dunnock is resident and sedentary, but Scandinavian and eastern birds are migratory. They winter in western and southern Europe and to a lesser extent in North Africa and parts of the Middle East. Though not gregarious at any time, the species is often seen in small parties when on migration and may be found in open, unvegetated habitats at this time, including sea-coasts.

Migration: Southward movements from September to November, returning in March–May.

Length:	15 cm
Wing length:	6.4–7.2 cm
Weight:	20 g
Voice:	'Tseet'. Song: thin high-pitched warble
Breeding period:	Mid-April, May, June. 2–3 broods per year. Replacement clutch possible
Size of clutch:	4–5 (–7) eggs
Colour of eggs:	Bright blue
Size of eggs:	19.5 × 14.5 mm
Incubation:	12–13 days, beg. from complete clutch
Fledging period:	Nidicolous; leaving nest at 13–14 days, but led a short time longer

ʼaria, 4.6.1978 (Pf

Young, Bavaria, June 1973 (Pf)

Bavaria, 4.5.1972 (Pf)

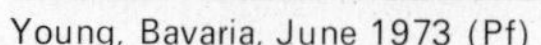

) Ad. at nest, Bavaria, 4.6.1978 (Pf)

Fan-tailed Warbler (Cisticola juncidis)

The Fan-tailed Warbler has an extensive world range, breeding in Africa, across Asia to the Far East and in Australia. It has recently extended its European range from the Mediterranean countries to the Channel coast of France.

It is found in reed-beds, marshland with juncus and other vegetation, rice fields, marshy meadowland, cereal crops and grassland. It feeds on small insects picked from the stems of plants or caught on the wing in short flights, skimming along the top of vegetation. The ♂ has a distinctive display-flight with bouncing undulations, sometimes quite high above the ground, uttering a sharp call, which is often the best indication of its presence.

The nest is well hidden in a clump of tall grass or rushes, sometimes only a few inches above the ground. It is a deep, purse-shaped structure, neatly made by weaving spiders' webs, dry grass, plant-down and grass flowers.

Both sexes share in incubation and rearing the young.

The Fan-tailed Warbler is mainly resident and sedentary in Europe. However, some movement does occur and the species has been recorded as a vagrant in Britain and Ireland and on passage to some Mediterranean watch-points. It is not known to what extent the species may move to North Africa, and the occurrence to the north of the breeding range are presumably connected with its extension of range.

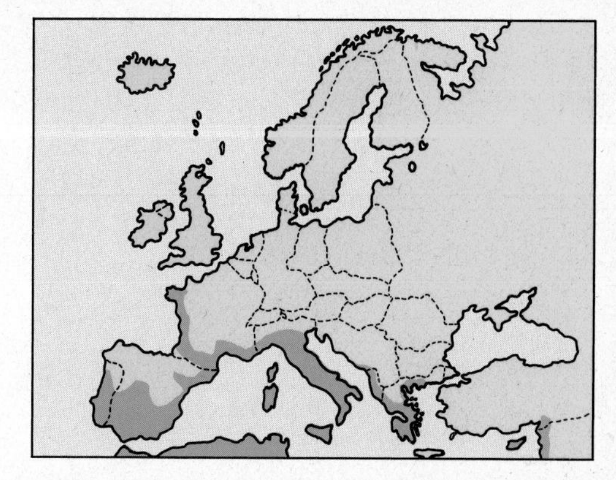

Length:	10 cm
Wing length:	4.9–5.2 cm
Weight:	6–8 g
Voice:	'Dsip-dsip-dsip', monotonously repeated in display-flight
Breeding period:	Beginning of April, May, June. 2 (rarely 3) broods per year
Size of clutch:	4–6 eggs
Colour of eggs:	Very variable. Bluish white with no spots, or thinly to densely spotted in pale red to black
Size of eggs:	15.5×11.5 mm
Incubation:	10 days, beg. from complete clutch
Fledging period:	Nidicolous; length of time spent by young in nest is not known

, Greece, 13.6.1975 (Sy)

Ad., Greece, 16.6.1975 (Sy)

Young, 9 days old, Greece, 20.6.1974 (Sy)

) Ad. at nest, Greece, 16.6.1975 (Sy)

Cetti's Warbler (Cettia cetti)

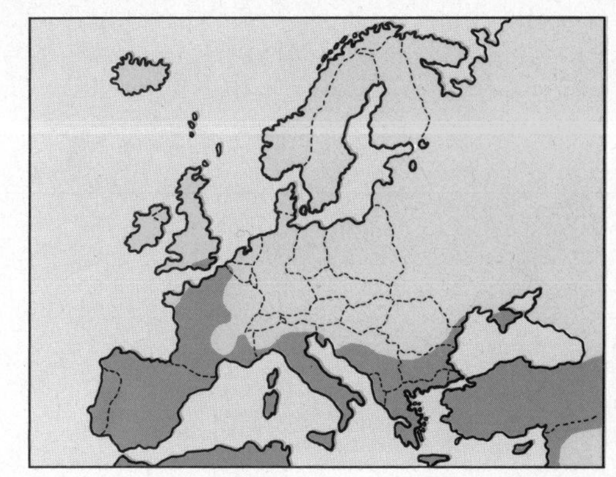

The Cetti's Warbler inhabits thickets and low bushes in swampy areas, riverside bushes, reed-beds with scrubby bushes and also hedges and areas of thick cover near cornfields, etc. It is a secretive bird and difficult to see, though the loud song gives its presence away.
It feeds on a variety of insects, mostly picked off plants and trees. Some earthworms, small molluscs and a few seeds are also taken.
The ♂ sings throughout the year though irregularly in the winter months. The sudden outburst of song is usually given from thick cover.
The nest-site may be in reeds or lush marsh vegetation, sometimes at ground-level, or in a bush or low tree up to 1 m from the ground. The nest is a large untidy cup of reeds, plant stems and similar material with a neater lining of fine grass, hair, feathers and reed-flowers. It is built by the ♀ who also incubates alone. Both sexes feed the young.
The Cetti's Warbler is mainly resident, though some dispersal may take place. It has extended its range into north-western Europe in recent years and now breeds regularly in much of southern England, having nested for the first time in 1972.
Migration: Most coastal migrants occur in October–November and March–April.

Length:	14 cm
Wing length:	5.4–6.6 cm
Weight:	13–15 g
Voice:	'Twik' or 'hyut'. Song: 'Chee-cheeweeoo-weechoo-weechoo'
Breeding period:	In west: End of April, May, early June. In east: Early May, June, early July. 1–2 broods per year
Size of clutch:	4–5 (3) eggs
Colour of eggs:	Variable, from dirty pink through chestnut brown to brick-red, almost always without spots
Size of eggs:	18.5×14 mm
Incubation:	13–14 days, beg. from complete clutch
Fledging period:	Nidicolous; length of stay in nest unknown

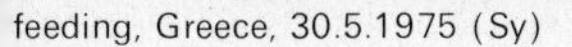

feeding, Greece, 30.5.1975 (Sy)

Derneburg, West Germany, 11.7.1975 (Becker)

) Ad. at nest, Greece, 30.5.1975 (Sy)

River Warbler (Locustella fluviatilis)

The River Warbler inhabits waterside areas, preferring dense undergrowth. It also occurs in swampy clearings in forest, bushy areas near city parks and even in cornfields. The habitat is as for Savi's Warbler and the two species behave as siblings in much of the range. Their diet is almost exclusively insects and their larvae.
The nest is a loose cup of plant stems, leaves and grass and is sited in a low shrub or on the ground in lush vegetation. The role of the sexes is incubation and nest-building is not known.
The River Warbler is a migrant, wintering in East Africa.
Migration: Departs breeding areas in August–September, returning in late April–June.

Savi's Warbler (*Locustella luscinoides*) The Savi's Warbler is also a secretive bird, though it is less terrestrial than the River Warbler. The ♂♂ often give their reeling song from an exposed reed-stem. The nest is a loose cup of reeds and marsh plants, sited on the ground in cover or in reeds about 1 foot above the water. It is built by the ♀, who incubates alone. She is fed by the ♂ when brooding. Both sexes feed the young, though mainly the ♀. The species is migratory, wintering in the Sudan, though little is known of its winter range.
Migration: Departs breeding areas in July–August, returning in April and May. The species became extinct as a breeding bird in Britain in the mid-1800s but recolonised from the 1960s.

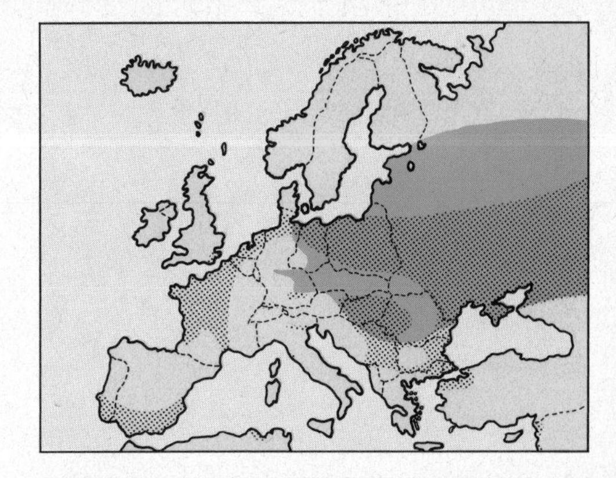

Length:	15 cm
Wing length:	6.7–7.8 cm
Weight:	16–20 g
Voice:	Reeling call, similar to Grasshopper Warbler, but slower and ending with 'swiswiswi'
Breeding period:	End of May, June, early July. 1 brood per year. Replacement clutch possible
Size of clutch:	4–5 (6) eggs
Colour of eggs:	White, with small and large grey-brown spots, denser towards the blunt pole
Size of eggs:	20.5×15 mm
Incubation:	13 days, beg. from complete clutch
Fledging period:	Nidicolous; leaving nest at 10–13 days

…r Warbler at nest, Lower Franconia, 16.6.1976 (Bosch)

River Warbler clutch, Austria, 1971 (Rei)

Savi's Warbler ♂, singing, Austria, May 1970 (Li)

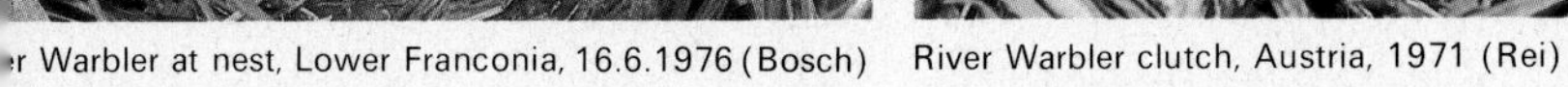

…) River Warbler ♂, singing, Lower Franconia, West Germany, 13.6.1976 (Bosch)

Grasshopper Warbler

(Locustella naevia)

The Grasshopper Warbler occurs in a variety of habitats; both dry and wet areas with tall grass, marshes, heath and moorland, young plantations, rough pastures and similar areas with a cover of low bushes, brambles, nettles, etc. It is a secretive species, keeping to cover most of the time and is very elusive, vanishing from a patch of cover without being seen. The song is distinctive, often given from an exposed perch, and it sings mainly in the early mornings and evenings but also at night. The diet is a variety of small insects and their larvae as well as spiders, small worms and woodlice.

The nest-site is on the ground or in low cover, usually with one entrance path for the birds. The nest is a cup of dead leaves, plant stems, dry grass and similar materials. Both sexes share in nest-building, incubation and rearing the young.

Grasshopper Warblers are entirely migratory. Very little is known of their winter quarters but some certainly winter in Ethiopia. No doubt the skulking habits and difficult habitat make it a difficult bird to find in winter.

Migration: Departs breeding areas in late July–September, adults leaving ahead of young birds. Return passage in April–May.

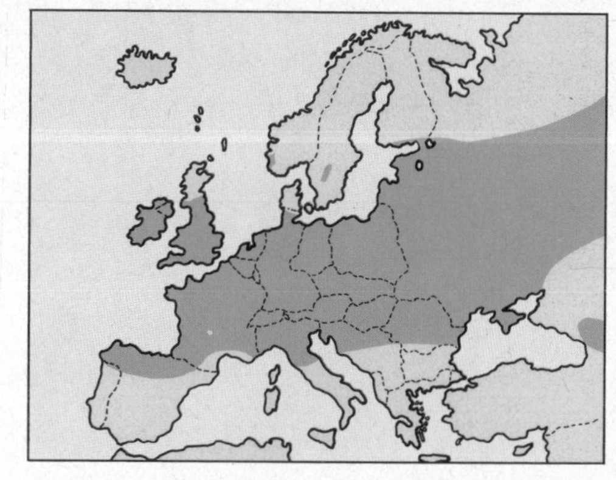

Length:	13 cm
Wing length:	5.7–6.6 cm
Weight:	13 g
Voice:	Song resembles that of line running off angler's reel, continuous reeling 'sirrr'. Call: 'tchik'
Breeding period:	End of April, May and June, July. 1–2 broods per year. Replacement clutch possible
Size of clutch:	6 (4–7) eggs
Colour of eggs:	White with numerous fine reddish-brown spots, often concentrated in a zone at the blunt pole
Size of eggs:	17.6×13.5 mm
Incubation:	14 days, beg. from complete clutch
Fledging period:	Nidicolous; leaving the nest at 10–12 days, still unable to fly and remaining in dense cover

at nest, Bavaria, 18.6.1975 (Pf)

Nestlings, 4 days old, Bavaria, 10.6.1976 (Pf)

Amperaue, Bavaria, 26.5.1976 (Li)

') Ad. at nest, Bavaria, 12.6.1974 (Pf)

Sedge Warbler

(Acrocephalus schoenobaenus)

The Sedge Warbler inhabits reed and osier beds with rank vegetation, riverside bushes, ditches with thick cover and sometimes bushy areas away from water. It feeds mostly on small insects and larvae as well as worms, slugs and a few berries. It creeps about in low vegetation, making only short flights. The nest-site is either on the ground in low cover or bound into a clump of reeds or plant stems. The nest is constructed with plant stems, grass and spiders' webs and is lined with fine grass, plant down and hair. Nest-building is carried out by the ♀ who also takes the major part of incubation. Both sexes feed the young. The Sedge Warbler is entirely migratory. The winter quarters are not fully known but are in Africa south of the Sahara. The species congregates at certain large reed-beds before migration and puts on a great amount of fat, sometimes reaching twice the summer weight before migrating.
Migration: August–September, returning in April–June.

Two similar species also breed in Europe, the **Aquatic Warbler** (*Acrocephalus paludicola*), which breeds mainly in eastern Europe and winters in Africa, and the **Moustached Warbler** (*Acrocephalus melanopogon*), which is found in Mediterranean countries. It is a partial migrant; some winter in the extreme south of Europe, others migrate as far as the Sudan. Both species occur in habitats similar to those of the Sedge Warbler and the ranges overlap in some areas.

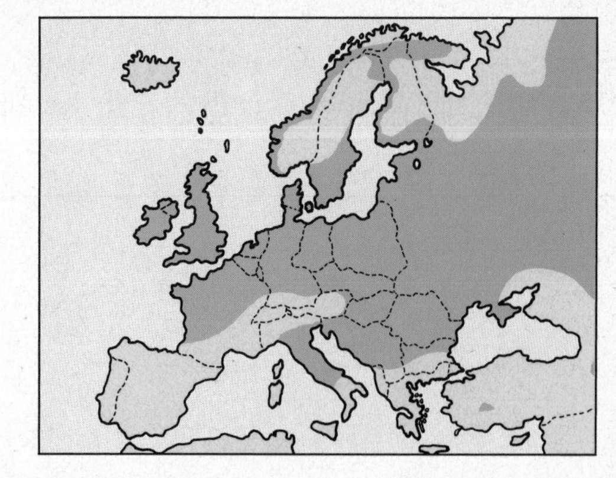

Length:	13 cm
Wing length:	5.9–7.2 cm
Weight:	12 g
Voice:	Song is loud and varied; musical phrases interspersed with mimics of other birds. Call: a repeated explosive 'tuk'
Breeding period:	Mid-May, June. 1 brood per year. Replacement clutch possible
Size of clutch:	5–6 (3–8) eggs
Colour of eggs:	Yellowish white, with fine and dense grey-brown to muddy spots, often concentrated in a zone at the blunt pole
Size of eggs:	17.5 × 13.0 mm
Incubation:	14 days, beg. from complete clutch
Fledging period:	Nidicolous: remaining 10–12 days in nest

vaiting to remove faecal sac, Seewinkel, Austria, 1968 (Li)

Fledged young bird, Turkey, 21.6.1977 (Li)

Seewinkel, Austria, 1970 (Li)

t) Seewinkel, Austria, 1968 (Li)

Marsh Warbler (Acrocephalus palustris)

The Marsh Warbler inhabits willow thickets, bushes and dense vegetation near lakes and river, usually with nettles, brambles or reeds, and sometimes hedgerows, cornfields and other agricultural habitats. It feeds mainly on small insects and larvae, particularly of marsh-dwelling species. Some spiders and small berries are also taken. The young are fed mainly on small caterpillars and midges. It picks insects from leaves and twigs in a manner similar to that of *Phylloscopus* warblers. Though it is sometimes rather secretive, it is less skulking than the Reed Warbler and often sings from an exposed perch.
The nest-site is usually 1–2 feet from the ground in tall vegetation, or up to 3 m from the ground in a bush. The nest is a deep cup, sometimes tapering towards the base with 'handles' which attach it to surrounding plant stems. It is built of dry grass, plant stems and similar material lined with fine grass, roots and hair. Both sexes take part in nest-building, though chiefly the ♀. The duties of incubation and rearing the young are shared.
The Marsh Warbler has extended its range into northern Europe in recent times. It is entirely migratory, wintering in east and south-east Africa.
Migration: Departs European range in late July–September, returning in late April–June.

The similar **Blyth's Reed Warbler** (*Acrocephalus dumetorum*) occurs in north-east Europe. It has a habitat preference similar to the Marsh Warbler and behaves as a sibling species in some areas. It winters in India, Ceylon and Burma.

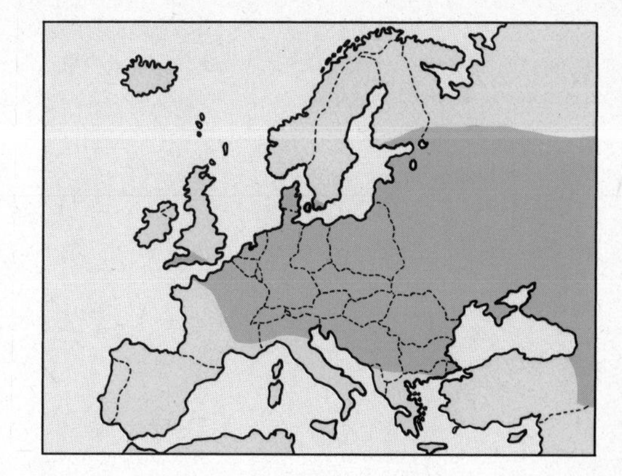

Length:	13 cm
Wing length:	6.3–7.0 cm
Weight:	12 g
Voice:	Extensive repertoire of fluting and canary-like trilling sounds, as well as imitations of other birds
Breeding period:	End of May, June, July. 1 brood per year. Replacement clutch possible
Size of clutch:	4–5 (3–6) eggs
Colour of eggs:	Bluish white, with fine spots of olive-brown, greenish and grey, often concentrated at the blunt pole
Size of eggs:	19×13.5 mm
Incubation:	13 days, beg. from complete clutch
Fledging period:	Nidicolous; remaining 10–13 days in nest

ged young bird, Bavaria, 25.6.1976 (Pf)

Nestlings, 4 days old, Bavaria, 19.6.1977 (Pf)

Bavaria, 28.6.1979 (Pf)

) Ad. at nest, Bavaria, 20.6.1976 (Pf)

Reed Warbler (Acrocephalus scirpaceus)

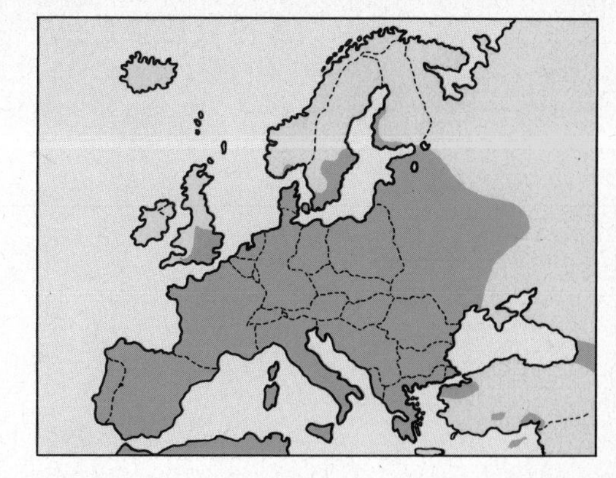

The Reed Warbler is normally found in reed-beds or marshy areas with reeds and other tall vegetation. It also occurs in cornfields and other cultivated land away from water. It is mainly associated with Phragmites reeds, however, and even a small clump of reeds in a ditch is sufficient for nesting. The diet is mainly small marsh-dwelling insects and larvae, and some spiders, worms, slugs and berries are also taken. It is generally a skulking species, keeping to cover where it searches for food. It does not take to the wing readily and flies for only short distances before dropping into cover. The nest-site is usually among the stems of growing reeds or other marsh vegetation. The nest is a deep cup of grasses, plant-stems, reed-flowers and spider's webs woven around upright stems. It is lined with fine plant material and sometimes hair or feathers. The nest is built mainly by the ♀, while both sexes share incubation and care of the young. The species often nests in loose colonies as a result of the preference for reed-beds. It is not, however, generally gregarious.
Reed Warblers are migratory, wintering in west, central and east Africa. European birds probably winter mainly in the west. Large numbers congregate in selective reed-beds prior to migration. Here they feed on aphids and other insects and may increase their body-weight considerably before departing for Africa.

Migration: Departs European breeding areas in August–September, returning in April–May.

Length:	13 cm
Wing length:	5.9–6.8 cm
Weight:	13 g
Voice:	'Check' and 'churr'. Song is series of chirping phrases: 'Chirri-chirri-trek-trek-tserr-tserr-tserr'
Breeding period:	End of May. June. 1 brood per year. Replacement clutch possible
Size of clutch:	4 (3–6) eggs
Colour of eggs:	Greenish white to blue-green, with irregular dark greenish-grey patches
Size of eggs:	18×13.4 mm
Incubation:	11 days, beg. from complete clutch
Fledging period:	Nidicolous; young remaining *ca* 2 weeks in nest

with young Cuckoo, Bavaria, 19.6.1975 (Pf)

Nestlings, 1 day old, Bavaria, 18.6.1976 (Pf)

Bavaria, 15.6.1976 (Pf)

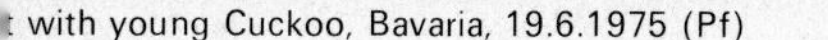

) Bavaria, 12.7.1977 (Pf)

Great Reed Warbler

(Acrocephalus arundinaceus)

The Great Reed Warbler is found exclusively in reed-beds in lakes, rivers and marshy areas and, except on migration, it is rarely seen away from its habitat. It feeds on insects and larvae, spiders, fresh-water shrimps and even small fish. Berries have been recorded as taken in autumn. Though it often keeps to the dense cover of reed-beds, it is not as skulking as the Reed Warbler, often perching on wires or bushes and frequently showing itself by short flights across the reeds or open water.

The nest-site is in reeds, the deep cup of plant stems, leaves, reed flowers being woven around reed-stems. The nest is built by the ♀ with the ♂ looking on. Both sexes share incubation and rearing the young. The species nests in loose colonies though nests are rarely close together. There is some competition with the Reed Warbler, which shares the same habitat, the larger Great Reed Warbler being dominant.

Great Reed Warblers have extended their range into northern Europe in recent times and are prone to overshooting on spring migration. Single birds have been heard singing in reed-beds well outside the normal range, including Britain.

The species is migratory, wintering over a large part of tropical Africa.

Migration: Departs Europe in August–September, returning in late April–June.

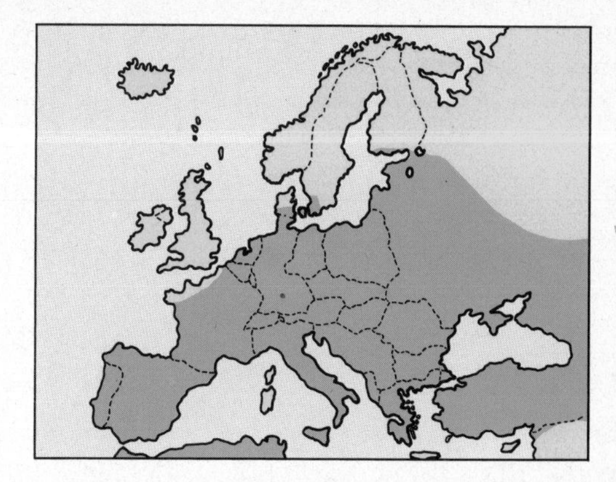

Length:	19 cm
Wing length:	8.8–10.0 cm
Weight:	30 g
Voice:	Harsh and grating 'Karra-karra-karra-kee-qurk-qurk-qurk'
Breeding period:	Mid-May, June. 1 brood per year. Replacement clutch possible
Size of clutch:	4–5 (3–6) eggs
Colour of eggs:	Bluish green to light blue, with rather dense coarse patches of olive to dark brown or blue-grey
Size of eggs:	22.5×16.3 mm
Incubation:	14 days, beg. from complete clutch
Fledging period:	Nidicolous; leaving the nest at 12 days, unable yet to fly, but climbing about in the reeds

ding ♀, Seewinkel, Austria, 1968 (Li)

Nestlings, Greece, 8.6.1975 (Sy)

Macedonia, Yugoslavia, 1.6.1979 (Li)

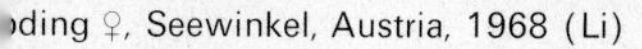

) Seewinkel, Austria, 1968 (Li)

Icterine Warbler (Hippolais icterina)

The Icterine Warbler frequents open broad-leaved woodland, parkland with scattered trees, orchards and gardens. It requires leafy trees, bushes and tall shrubs, where it spends most of its time, rarely descending to the ground or low cover. It feeds on a variety of insects and larvae, spiders, berries and ripe fruit. Though it is active and lively when feeding, it is often hard to see, as it remains in leafy cover much of the time.

The nest is usually in the fork of branches in a small tree or bush, sometimes in taller trees up to 4 m above ground. The nest is a deep cup usually bound into twigs or branches. It is made of plant stems, leaves and roots with a lining of fine grass, hair and sometimes feathers. Both sexes share nest-building, incubation and feeding the young.

The Icterine Warbler is entirely migratory, wintering in central Southern Africa.

Migration: Departs European breeding grounds in August–September, returning in late April to June.

In south-west Europe it is replaced by the **Melodious Warbler** (*Hippolais polyglotta*), which shares a similar habitat though it is found in lower vegetation than is the Icterine. The nest and breeding behaviour are very similar to those of the Icterine Warbler though incubation is by the ♀ alone. It is a migratory bird, wintering in tropical West Africa. It migrates in August–October and April–May.

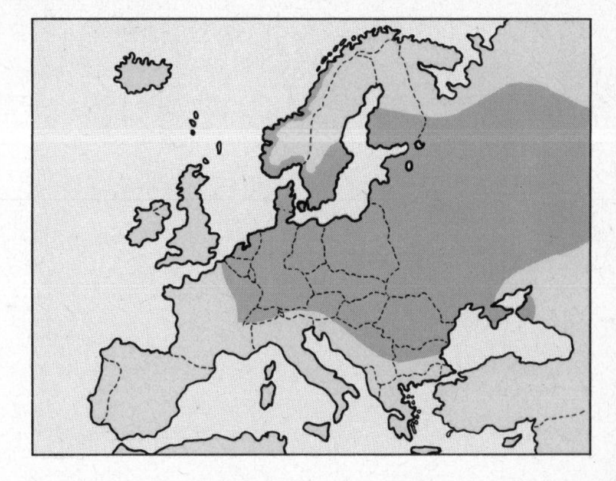

Length:	13 cm
Wing length:	7.1–8.3 cm
Weight:	13 g
Voice:	'Tek-tek' or 'dederoi'. Warning call: 'Err'. Song: mixture of trilling and chattering sounds
Breeding period:	Mid-May, June, July. 1 brood per year, rarely 2
Size of clutch:	4–5 (6) eggs
Colour of eggs:	Pink to violet, a few dark patches
Size of eggs:	18×13.5 mm
Incubation:	13 days, beg. from complete clutch
Fledging period:	Nidicolous; leaving the nest at 2 weeks

aria, 28.6.1978 (Pf)

Nestlings, Bavaria, 30.6.1978 (Pf)

Amperaue, Bavaria, 1975 (Li)

) Ad. at nest, Bavaria, 18.7.1979 (Pf)

Olive-tree Warbler

(Hippolais olivetorum)

The Olive-Tree Warbler has a restricted range in south-east Europe and parts of the Middle East. It occurs in oak-woods, olive groves, orchards and other broad-leaved trees in open areas. It is a secretive bird, usually skulking in leafy cover, and is difficult to observe. It feeds on a variety of insects, larvae, berries and perhaps some fruit. It is the largest of the *Hippolais* Warblers. The nest-site is in a bush or tree, often near the ground but sometimes as high as 3 m. The nest is a deep cup of plant stems, leaves, grasses and spiders' webs, lined with fine grass, hair or a few feathers. Little information is known about the role of the sexes in nest-building, incubation and care of the young. It is a migratory species, wintering in east and south-east Africa, usually in association with acacia trees.

Migration: Autumn movements in September–October, returning in late March–April.

The smallest species of *Hippolais*, the **Booted Warbler** (*Hippolais caligata*), occurs as a breeding bird in extreme north-east Europe. It is found in low birch woodland, willow thickets, riverside vegetation, cornfields and other open habitats. It feeds on small insects searching for food in low cover. It is secretive and skulking, behaving like an *Acrocephalus* Warbler. The nest is usually at ground-level amongst tall plants. Little information on breeding biology is available, though incubation is mainly by ♀. The species winters in Asia from Ceylon to Arabia. It migrates in August–October and May–June.

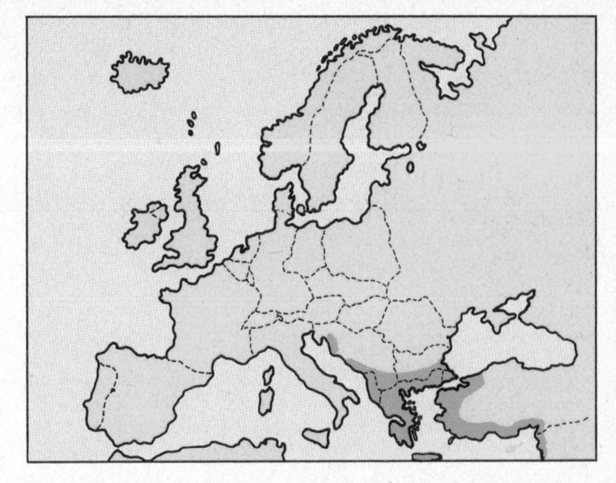

Length:	15 cm
Wing length:	8.2–9.0 cm
Weight:	*ca* 15 g
Voice:	Song louder, slower and deeper than other *Hippolais*
Breeding period:	Beginning of June. 1 brood per year. Replacement clutch possible
Size of clutch:	3–4 eggs
Colour of eggs:	Pale violet in various hues with a few blackish, irregularly distributed patches
Size of eggs:	21 × 14 mm
Incubation:	13 days, beg. from complete clutch
Fledging period:	Nidicolous; leaving the nest, not yet quite able to fly, at 2 weeks

Nestlings, Greece, 1.7.1977 (Li)

Greece, 30.5.1978 (Li)

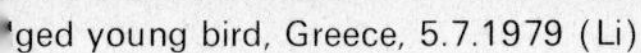

ged young bird, Greece, 5.7.1979 (Li)

) Ad. ♀, Greece, 2.6.1965 (Fe)

Olivaceous Warbler (Hippolais pallida)

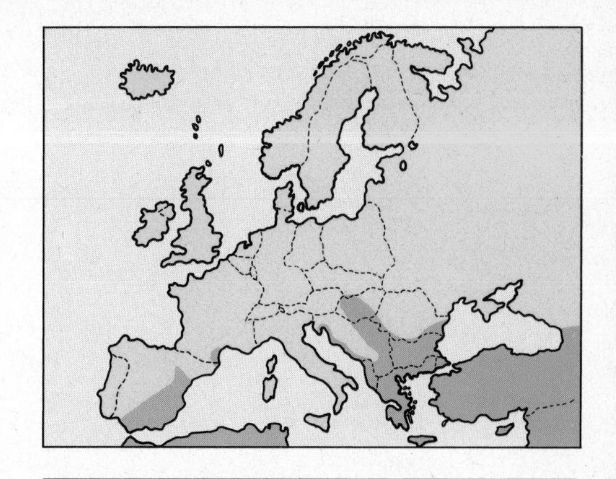

The Olivaceous Warbler is found in a wide variety of habitats including bushes in semi-desert or steppe areas to orchards, gardens, riverine bushes, thickets and scattered broad-leaved trees. In some areas it shows a preference for vegetation near water yet it also occurs in quite arid areas.
It feeds mainly on small insects and larvae, picked from amongst flowers and leaves. It also takes some berries in autumn. It usually keeps to thick cover and can be very difficult to observe. The scratchy song is similar to that of *Acrocephalus* Warblers and is usually delivered from dense vegetation. The nest is placed in the outer twigs of a bush and is a deep cup of small twigs, grasses, plant stems and similar material. It is lined with fine plant fibres and hair. Incubation is by the ♀ alone and both sexes care for the young.
There are two races of Olivaceous Warblers breeding in Europe; *opaca*, which breeds in Spain, and the smaller *eliaca*, in the Balkans and Asia Minor. The species is migratory, with *opaca* wintering in West Africa and *eliaca* in East Africa. They occupy different habitats at this season, *opaca* preferring dense Acacia and gardens and *eliaca* being found in mangroves and savanna trees.
Migration: Departs breeding areas in August–September, returning in April–May.

Length:	13 cm
Wing length:	6.3–7.2 g
Weight:	12 g
Voice:	Call: 'Tek tek'. Song: similar to Reed Warbler, but faster and more monotonous
Breeding period:	End of April, May, June. 2 broods per year. Replacement clutches follow very quickly
Size of clutch:	3–4 (5) eggs
Colour of eggs:	Dirty white, with a few black spots
Size of eggs:	17.5×13.4 mm
Incubation:	13 days, beg. from complete clutch
Fledging period:	Nidicolous; remaining 14 days in nest

ost-fledged young bird, 11.6.1976 (Li)

Greece, 21.6.1976 (Li)

Greece, 1.6.1978 (Li)

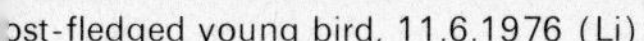

Ad. ♂, Yugoslavia, 1971 (Li)

Garden Warbler (Sylvia borin)

The Garden Warbler is widely distributed throughout much of Europe. It occurs in most types of woodland but shows a preference for broad-leaved or mixed forest with a dense undergrowth of bushes. It is also found in scrub, gardens, parks, tall hedgerows and bushy areas. It feeds on a variety of insects, spiders, small worms and beetles. In autumn it takes soft fruit and berries, mainly blackberry, elder, honeysuckle, etc. It is a quite unobtrusive bird, keeping to low cover most of the time.

The nest is placed in a low bush, tall vegetation or small tree usually between 1 foot and 9 feet above the ground. The nest is built of grasses, moss, fine twigs and roots lined with finer plant material. It is usually wedged in a fork of twigs or plant stems.

Nest-building, incubation and care of the young is shared by both parents. Garden Warblers are entirely migratory, wintering in much of Africa south of 10° N.

Migration: Departs breeding areas in mid-July to October, returning in April–June.

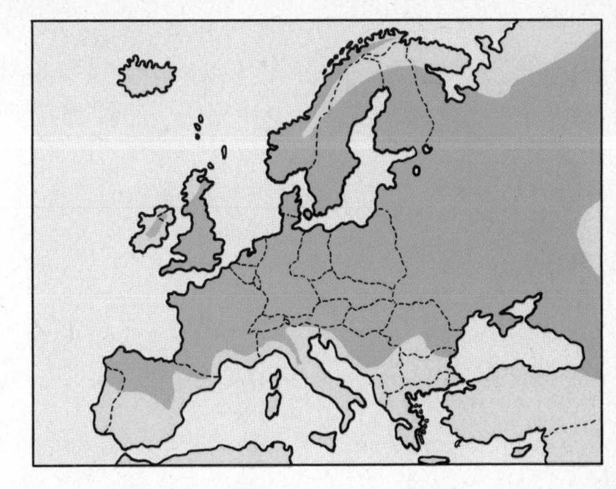

Length:	14 cm
Wing length:	7.3–8.4 cm
Weight:	20 g
Voice:	Song: sweet melodious warble. Call: a snapping 'tacc tacc'; also 'churr'
Breeding period:	Mid-May, June, July. 1 or 2 broods per year
Size of clutch:	4–5 (3–7) eggs
Colour of eggs:	Very variable. Greenish white with coarse and fine brownish patches of various hues
Size of eggs:	20 × 15 mm
Incubation:	12 days, beg. from complete clutch
Fledging period:	Nidicolous; leaving the nest at 9–10 days, not yet able to fly

. brooding, July 1973 (Pf)

Fledged young bird, Bavaria, 20.6.1976 (Pf)

Bavaria, 17.6.1977 (Pf)

ft) Ad., Bavaria, June 1977 (Pf)

Blackcap (Sylvia atricapilla)

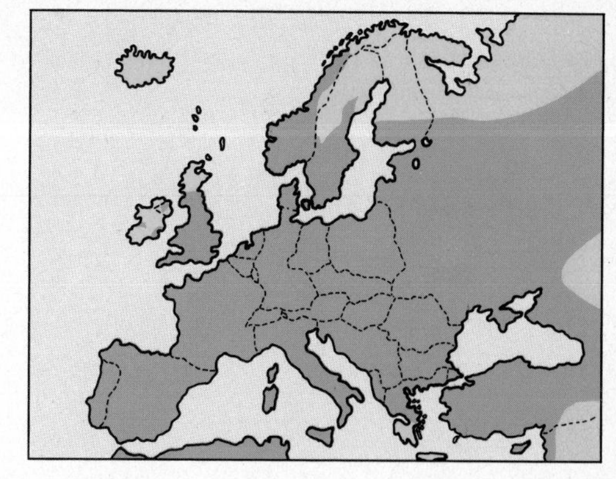

The Blackcap has a similar habitat preference to that of the Garden Warbler but shows a greater liking for coniferous woods. It is found in most types of woodland with a tall but not necessarily dense understorey, thickets, copses, hedgerows, riverine vegetation, parks and large gardens. It feeds on a wide variety of insects and larvae including small beetles. It also takes a large amount of berries and soft fruits in summer and autumn. In southern Europe it has been recorded feeding on oranges and figs.
It is an active and lively bird and, though it generally keeps to the cover of bushes and trees, it is less skulking than the Garden Warbler.
The usual nest-site is in a low tree or bush, usually higher than the nests of the Garden Warbler. The nest is a neat cup of dry grass, roots, moss, etc. lined with fine plant materials and hair. Both sexes share in nest-building and incubation.
The Blackcap is a partial migrant in Europe. Some birds winter as far north as Britain and through western Europe to Mediterranean countries, where it is probably resident. Most birds move to tropical Africa, where it occurs in the west and east of the continent. There is a migratory divide, with the more eastern breeding birds migrating down the east of the Mediterranean, thence to that side of Africa.
Migration: Departs northern and eastern breeding areas from August to October, returning in late March to May.

Length:	14 cm
Wing length:	6.7–7.8 cm
Weight:	20 g
Voice:	Call: harsh 'tacc tacc'. Song: rich warbling, more musical than Garden Warbler
Breeding period:	Mid-May, June, July. Usually 2 broods per year. Replacement clutch possible
Size of clutch:	4–5 (3–6) eggs
Colour of eggs:	Variable. Light brown or greenish white with dark brown and grey patches, often very densely marked
Size of eggs:	19.5 × 14.7 mm
Incubation:	12–14 days, beg. from 2nd or 3rd egg
Fledging period:	Nidicolous; leaving the nest at 12 days, unable yet to fly

♀, Bavaria, 15.5.1979 (Li)

Nestlings, Bavaria, 18.7.1978 (Pf)

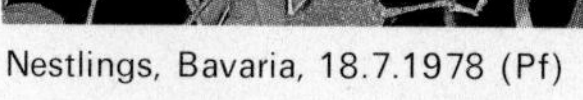

Bavaria, 18.5.1977 (Pf)

) Ad. ♂ feeding young, 22.6.1979 (Pf)

Lesser Whitethroat (Sylvia curruca)

The Lesser Whitethroat inhabits shrubby vegetation, woodland edges, hedgerows, gardens and parkland. It prefers dense, dark undergrowth with taller bushes and shrubs. A desert-form exists which occurs in sparse vegetation in arid regions.
The diet is mainly the larvae and eggs of insects, mainly Lepidoptera, as well as ants, spiders, small worms and beetles. In autumn it takes berries and soft fruit such as Blackberry and Elder.
It is a skulking bird, keeping to thick cover much of the time. It is not confined to low vegetation and will search for food in the tops of trees.
The nest-site is usually about 1 m from the ground in a bush or conifer; sometimes it may be as high as 3 m. The nest is a cup built of small twigs, dry grasses, leaves and roots with spider cocoons on the outside. It is lined with hair, fine roots and plant down. Both sexes share nest-building, incubation and rearing the young.
The Lesser Whitethroat is entirely migratory, wintering in north-east Africa. It has an unusual migration route for a species breeding so far west. In autumn the western population passes through Italy and the eastern Mediterranean, returning even further east, through the Levant in spring.

Migration: Departs breeding areas from mid-July with main passage in August–September continuing into October. It returns in April and May.

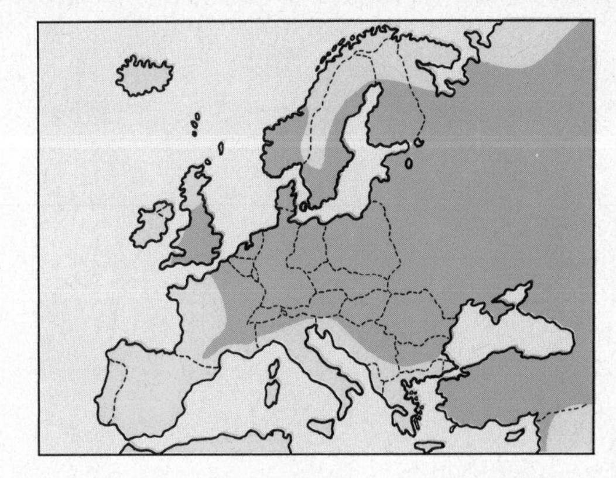

Length:	13 cm
Wing length:	6.1–6.8 cm
Weight:	12 g
Voice:	Song: rattling repetition of 'chikka-chikka-chikka-chikka-chik' and musical warbling. Call: 'tzek'
Breeding period:	Beginning of May, June. 1–2 broods per year. Replacement clutch possible
Size of clutch:	5–6 (3–7) eggs
Colour of eggs:	Whitish, patched in various shades of brown and grey, denser at the blunt pole
Size of eggs:	17.6 × 13.1 mm
Incubation:	12 days, beg. from complete clutch
Fledging period:	Nidicolous; leaving nest at 12 days, not yet able to fly

brooding, 14.7.1978 (Pf)

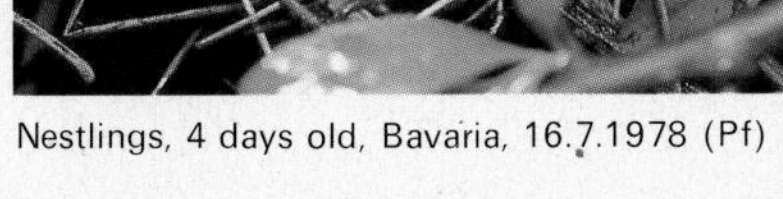

Nestlings, 4 days old, Bavaria, 16.7.1978 (Pf)

Turkey, 31.5.1973 (Li)

) Ad. at nest, Bavaria, 14.7.1978 (Pf)

Whitethroat (Sylvia communis)

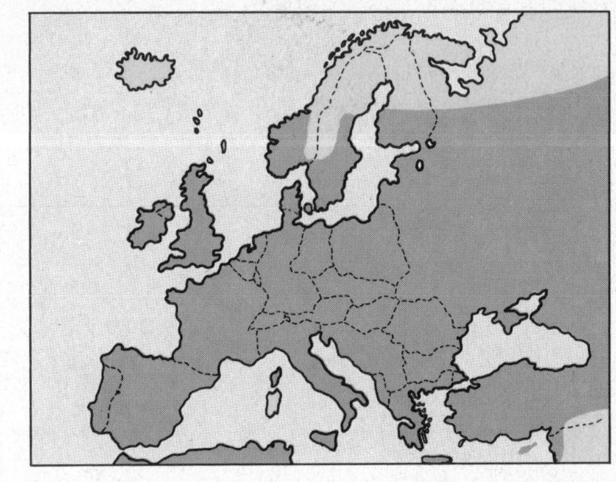

The Whitethroat occurs in a wide range of scrubby growth, hedges, roadside vegetation, forest edges, heathland and similar areas with low bushy cover. The diet includes a variety of insects and larvae, ants, beetles and spiders and in autumn a range of berries and soft fruit. The young have been recorded as fed on Cuckoo-spit and green caterpillars.

Though it sometimes keeps to thick cover, it is generally less skulking than most Warblers, moving through low vegetation with great agility. The flight is weak and jerky, ending with a dive into cover.

The nest-site is in low vegetation, usually close to the ground but sometimes as high as 5 m. The nest is a loose cup of grass, roots and similar material, lined with hair or fine roots. The ♂ often builds the nest before the arrival of the ♀, but sometimes the ♀ will build her own nest. Incubation and care of the young is shared by both parents.

The Whitethroat is entirely migratory, wintering in a belt across Africa south of the Sahara, the 'Sahel' and in east Africa. In 1968 there was a dramatic reduction in the population of Whitethroats from Britain and other areas in north-west Europe. This was probably due to prolonged drought conditions in the wintering area of the Sahel. The species has not recovered from this disaster, and populations are still at a low level in some areas.

Migration: Autumn passage from mid-July to October, mainly August–September, returning in late March to June.

Length:	14 cm
Wing length:	6.5–7.5 cm
Weight:	14 g
Voice:	Call: 'tacc-tacc', 'charr' and 'wheet wheet wit wit wit'. Song: a short rapid warble, rather scratchy
Breeding period:	Beginning of May, June, July. Usually 2 broods per year. Replacement clutch possible
Size of clutch:	5–6 (3–4) eggs
Colour of eggs:	Greenish white to light yellowish-brown, more or less densely flecked with shades of brown
Size of eggs:	18.5×13.9 mm
Incubation:	12 days, beg. from complete clutch
Fledging period:	Nidicolous, remaining 10–11 days in nest

avaria, 19.6.1978 (Pf)

Nestlings, 2 days old, Bavaria, 19.6.1978 (Pf)

Bavaria, 16.5.1976 (Li)

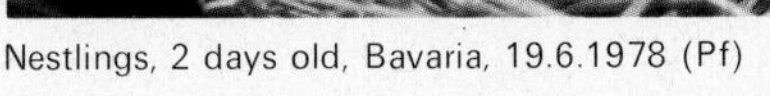

) ♂ at nest, Bavaria, 19.6.1978 (Pf)

Dartford Warbler (Sylvia undata)

The Dartford Warbler inhabits *maquis* and dry Mediterranean scrub. In the north of its range it is found in heathland with gorse and heather. As it is a resident bird, it is found only in mild or warm climates where there is sufficient insect food to sustain the population during the winter. In the extremes of the range, such as southern England, it is sometimes badly affected by hard winters, with the population being much reduced.
It feeds mainly on small insects including beetles, lepidoptera and larvae and spiders which are an important part of the diet in winter. It will also take small berries and fruit in autumn. It is a very secretive, skulking bird, keeping to dense cover. It has a weak, jerky flight and dives into fresh cover when distributed. In autumn and winter it often forms small parties which move around together.
The nest-site is in a low shrub such as gorse, bramble or cistus, often very close to the ground. The nest is a cup of plant material such as grass and roots with some moss, plant-down and spider-cocoons. It is lined with hair and fine plant matter. The ♂ makes flimsy 'cock's nests' but the ♀ is chiefly responsible for nest-building and incubation.
The Dartford Warbler is mainly resident and sedentary, though some dispersal takes place in autumn.
The similar **Marmora's Warbler** (*Sylvia sarda*) occurs in less dense scrub and *maquis* usually near the sea. It appears to have habits and behaviour very similar to the Dartford Warbler. It is mainly resident but occurs in North Africa and elsewhere in winter, suggesting some movement away from breeding areas.

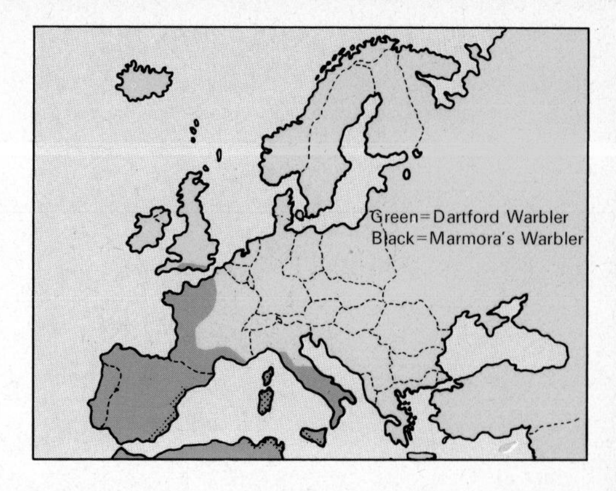

Length:	13 cm Dartford Warbler
Wing length:	4.8–5.7 cm
Weight:	8.5 g
Voice:	Song resembles that of Whitethroat; rattling alarm call, 'tchirr'
Breeding period:	Mid-April, May, June in the north; early April, May, June in the south. 2–3 broods per year
Size of clutch:	3–4 (–6) eggs
Colour of eggs:	Whitish to pale greenish, with olive or red-brown patches of very variable density and size
Size of eggs:	18×14 mm
Incubation:	12–13 days, beg. from complete clutch
Fledging period:	Nidicolous, remaining 11–12 days in nest

tlings, Spain, June 1980 (Sy)

Marmora's Warbler ♂, captive bird, Radolfzell (Wü)

Length:	12 cm Marmora's Warbler
Wing length:	5.0–5.8 cm
Weight:	*ca* 8 g
Voice:	Song resembles Dartford Warbler's, but less harsh. Call: 'Tsiq'
Breeding period:	Mid-April, May, June. 2 broods per year
Size of clutch:	3–4 eggs
Colour of eggs:	Whitish to pale greenish with olive to brown-grey patches, always denser at the blunt pole
Size of eggs:	18×13.4 mm
Incubation:	11–12 days, beg. from complete clutch
Fledging period:	Nidicolous; no information

) Dartford Warbler ♂, Spain, June 1980 (Sy)

Subalpine Warbler (Sylvia cantillans)

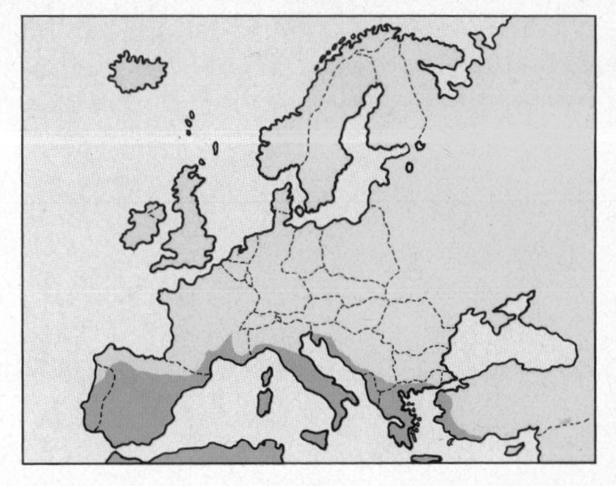

The Subalpine Warbler is found in Mediterranean scrub, open woodland (particularly oak with a dense undergrowth) and sparse gorse scrub. It inhabits the same areas as Dartford and Spectacled Warblers but can breed at a much higher altitude than those species, reaching 2,000 m in some areas. It is often found on sunny mountain slopes. The diet is mainly small insects and larvae, spiders and a few seeds. It is generally very skulking and keeps to cover. The nest-site is in a low shrub or other vegetation rarely more than 1 m above ground. The nest is a cup built of grass, plant fibres and down lined with finer plant material. Incubation is mainly by the ♀ with both parents feeding the young.

The Subalpine Warbler is a migrant, wintering in a belt across tropical West Africa.

Migration: Departs breeding areas in September–October, returning in late March to May.

The **Spectacled Warbler** (*Sylvia conspicillata*), which resembles a small, bright Whitethroat, occurs in a habitat similar to that of the Subalpine Warbler in Mediterranean countries. It prefers a lower habitat to the Subalpine Warbler, such as coastal plains. It is partly resident, wintering birds being recorded in southern France and Iberia, but some migrate to North Africa.

Length:	12 cm
Wing length:	5.3–6.2 cm
Weight:	10 g
Voice:	'Chat-chat-chat' and 'tecc'. Song resembles Whitethroat, but more musical
Breeding period:	April, May, June. 2 broods per year
Size of clutch:	3–4 (5) eggs
Colour of eggs:	Greenish white, with dense and fine dark brown to grey-violet spots, concentrated at the blunt pole
Size of eggs:	17×13 mm
Incubation:	12 days, beg. from complete clutch
Fledging period:	Nidicolous; remaining 12 days in nest, but not then fully able to fly

nest, Greece, 7.6.1976 (Li)

Fledged young bird, Greece, 11.6.1976 (Li)

Greece, 6.6.1976 (Li)

) Ad. ♂, Greece, 7.6.1976 (Li)

Rüppell's Warbler (Sylvia rueppelli)

The Rüppell's Warbler has a very restricted range in extreme south-east Europe and south-west Asia. It is found in thick scrub, often thorny bushes in dry rocky gullies and hillsides. It also occurs in mature open woods of oak and cypress. It breeds in mountainous areas to a height of 1,500 m as well as in lower rocky areas. The habitat overlaps with that of the Sardinian Warbler in some areas. The diet is mainly insects and larvae with some berries and fruit in autumn. The young have been recorded as fed on green caterpillars. It is generally a secretive bird, keeping to cover, though the ♂♂ often perch on exposed places when singing. It does not keep exclusively to low cover and is often seen in trees.

The nest is sited in a bush or small tree and is a neat cup of grasses, plant stems and similar material, often with spider cocoons on the outside. The ♂♂ may make several 'cock's nests' and one of these may be used or a fresh one built by both sexes. Incubation is shared but is chiefly done by the ♀.

The Rüppell's Warbler is entirely migratory, wintering in very arid country in the Sudan and neighbouring countries in Africa.

Migration: Departs breeding areas in August–September, returning in late February to March.

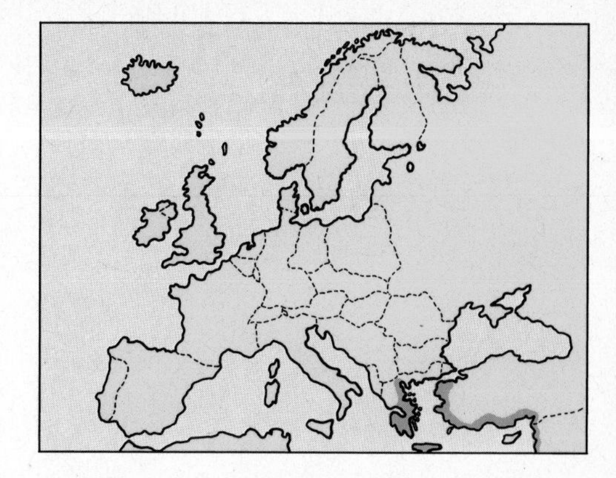

Length:	14 cm
Wing length:	6.4–7.4 cm
Weight:	12 g
Voice:	Rattling warning call. Song: Whitethroat-like chatter with rattling sounds
Breeding period:	Mid-April, May. 1 brood per year
Size of clutch:	5 (4–6) eggs
Colour of eggs:	Whitish green to grey, with dense olive-brown or yellowish spots, often concentrated at the blunt pole
Size of eggs:	18×13.6 mm
Incubation:	13 days, beg. from complete clutch
Fledging period:	Nidicolous; further information not known

vith dark head-colouring), Greece, 6.5.1974 (Fe)

♀ (typical) with young, Greece, 4.5.1975 (Fe)

Greece, 23.4.1974 (Fe)

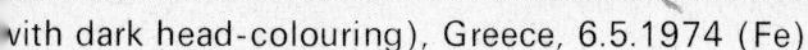

) Ad. ♂ at nest, Greece, 12.5.1974 (Fe)

Sardinian Warbler (Sylvia melanocephala)

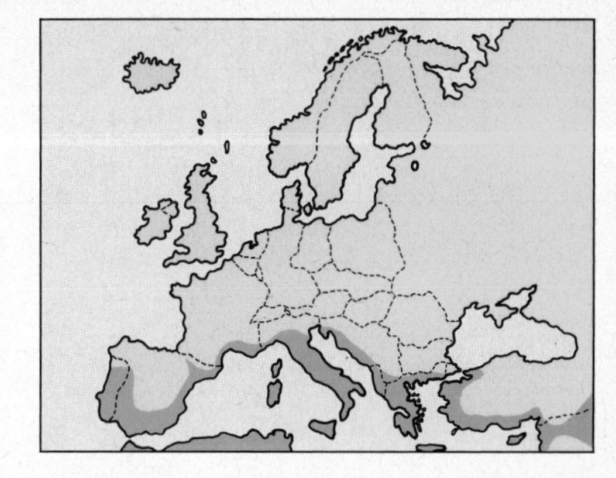

The Sardinian Warbler is a typical inhabitant of Mediterranean *maquis* scrub, often with taller trees and bushes. It also occurs in thickers, hedgerows and areas of low cover in gardens and parks. It shares some habitat with Subalpine and Dartford Warblers and there may be some inter-specific competition with these species. The Sardinian Warbler feeds mainly on small insects and larvae, and in autumn it takes berries and soft fruit. Some seeds are taken in winter.

It is a secretive bird, keeping to low cover much of the time, and often the only clue to its presence is the characteristic call. The nest-site is usually a short distance from the ground in brambles, scrub or tall vegetation. The nest is a cup of grass, plant stems and similar material with spiders' webs used to bind the structure. It is lined with fine plant material and hair. Incubation and rearing the young is shared by both parents with the ♀ taking the larger share. The Sardinian Warbler is mainly resident and the population may suffer badly from the effect of hard winters in the north of its range. There is some movement to North Africa and some birds occur south of the Sahara in both West and East Africa.

Length:	13.5 cm
Wing length:	5.3–6.2 cm
Weight:	10–15 g
Voice:	Rattling warning call: 'cha-cha-cha-cha'; also 'trek'. Song: resembles Whitethroat's, with rattling notes
Breeding period:	Mid-March, April, May, June, July. 2 broods per year
Size of clutch:	4–5 (3) eggs
Colour of eggs:	Light green, white or reddish with dense grey, brown or chestnut spots
Size of eggs:	18×13.5 mm
Incubation:	13–14 days
Fledging period:	Nidicolous; leaving the nest at 12 days, not yet able to fly

♀, Greece, 8.6.1976 (Li)

Greece, 7.6.1976 (Li)

Greece, 14.6.1979 (Li)

Ad. ♂, Greece, 8.6.1976 (Li)

Orphean Warbler (Sylvia hortensis)

One of the largest of the *Sylvia* warblers, the Orphean is found in open woodland including oak and conifers with little or no understorey. It also inhabits olive groves, citrus plantations, dry scrub or hilly areas, gardens and parks.
It feeds on a variety of insects and larvae, mostly Lepidoptera, as well as fruit and berries in autumn and winter. It is mainly arboreal, searching for food in leafy trees, though frequently descends to low cover at times.
The nest-site is usually in the outer branches of trees or bushes 1–2 m from the ground. The nest is a loose cup of twigs, plant stems, grasses and roots, often with spiders' webs and cocoons on the rim. It is lined with fine plant material and hair. Incubation and care of the young is shared by both sexes, with the ♀ taking the main part in incubation.
The Orphean Warbler is entirely migratory, wintering in a narrow belt across Africa south of the Sahara, where it occurs in acacia savanna and palm groves. The European range has gradually contracted southwards since the last century, probably due to climatic change as the species prefers warm regions.
Migration: Departs breeding areas in September–October, returning from late March to April.

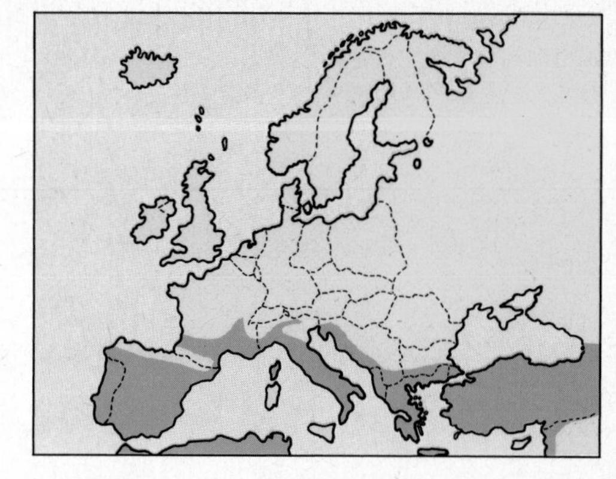

Length:	15 cm
Wing length:	7.2–8.3 cm
Weight:	22–24 g
Voice:	'Tsek, tsek' and rattling 'terr'. Song: melodious, thrush-like warble, sometimes imitative of other birds
Breeding period:	Beginning of May, June. 1 brood per year. Replacement clutch possible
Size of clutch:	4–5 (6) eggs
Colour of eggs:	Greenish white, with sparse patches of blue-grey and different shades of brown, often concentrated at the blunt pole
Size of eggs:	19.6×14.6 mm
Incubation:	13 days, beg. from complete clutch
Fledging period:	Nidicolous; remaining 12–14 days in nest

♂, Greece, 2.6.1978 (Li)

Greece, 2.6.1978 (Li)

Greece, 29.5.1976 (Li)

) Ad. ♀, Greece, 13.6.1976 (Li)

Barred Warbler (Sylvia nisoria)

The barred Warbler inhabits thickets of riverine woodlands, damp boggy areas, hedgerows, forest edges, orchards and parkland. It often shows a preference for thorny bushes and in some places the habitat overlaps with Whitethroat and Red-backed Shrikes.
It feeds on insects and larvae including beetles and Lepidoptera, and in autumn and winter it takes berries and soft fruit as well as worms. It is a rather skulking bird, moving easily through dense cover, and is difficult to observe.
The nest-site is a bush or small tree, usually in a fork of branches or twigs. The nest is a large, deep cup of grass and plant stems, decorated with spiders' cocoons on the rim. It is lined with hair and fine plant material. Both sexes share incubation and care of the young.
The Barred Warbler is entirely migratory, wintering in thorn scrub in East Africa and Arabia. It normally migrates south-east in autumn but a percentage of juvenile birds appear to reverse this direction, and fly north-west to make a landfall in Britain, particularly in the northern isles. There is evidence that at least some of these birds are able to re-orientate and return to south-east Europe.
Migration: Departs breeding areas in August–September, returning in late April to early June.

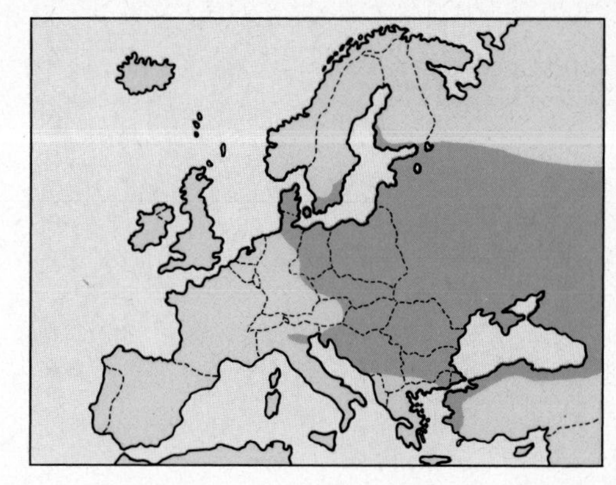

Length:	15–16 cm
Wing length:	8.3–9 cm
Weight:	30 g
Voice:	Calls: 'charr-charr-charr' and harsh 'chack-chack'. Song: resembles Garden Warbler with more rattling sounds
Breeding period:	End of May, June. 1 brood per year. Replacement clutch possible
Size of clutch:	5 (3–6) eggs
Colour of eggs:	Grey-white with many light blue-grey patches and murky coloration, generally concentrated at the blunt pole
Size of eggs:	21×15 mm
Incubation:	14 days, beg. from complete clutch
Fledging period:	Nidicolous; leaving the nest at 15 days, not yet fully able to fly

♂, Hansag, Austria, 1968 (Li)

Hansag, Austria, 1968 (Li)

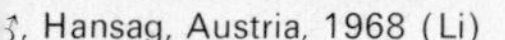

♀, Seewinkel, Austria, 1968 (Li)

Chiffchaff (Phylloscopus collybita)

The Chiffchaff is one of the most widespread warblers in Europe, with separate races in Iberia, Scandinavia and eastern Europe and in the extreme north-east of Europe. it occurs in a variety of habitat, mostly open woodland, both broad-leaved and coniferous, as well as areas with trees in parkland, gardens, roadsides, etc. In mountain areas it occurs to a height of 2,000 m in oak and pine woods. It feeds mainly on small insects, particularly midges and the larvae of Lepidoptera, and spiders. Most of the food is collected in the foliage of trees and bushes, where it flits about actively, fluttering and hovering.

The nest is a slightly domed cup of moss, plant débris and leaves lined with feathers. It is placed in the ground in tall vegetation or in low bushes, tree branches or creepers. It is built entirely by the ♀, who also incubates alone, leaving only for a short period to feed. The young are normally fed by the ♀ only, though sometimes the ♂ may assist.

The Chiffchaff is mainly migratory, wintering from the Mediterranean countries to North Africa and in tropical Africa south of the Sahara. In southern Europe it is partly resident. Some birds winter as far north as Britain and Ireland; these are not local residents but birds from Scandinavia or further east.

Migration: Departs breeding areas from August to October, returning in March–May.

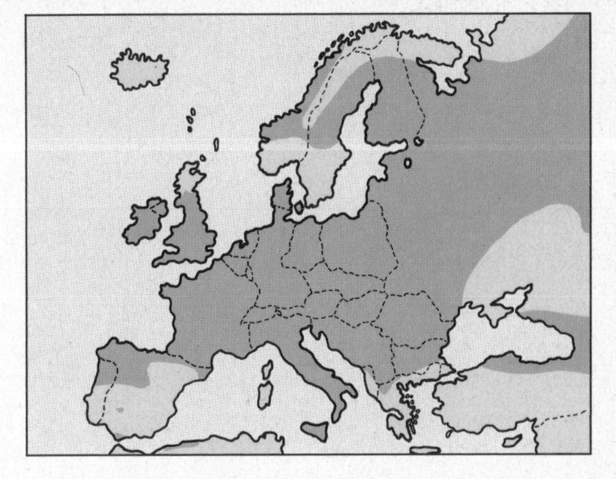

Length:	11 cm
Wing length:	5.3–6.3 cm
Weight:	9 g
Voice:	Calls: 'hweet' and 'twit'. Song: 'chiff-chaff, chiff-chaff', monotonously repeated
Breeding period:	End of April, May, June, mid-July. 1 or 2 broods per year. Replacement clutch possible
Size of clutch:	5–6 (4–7) eggs
Colour of eggs:	Whitish, with sparse purple-brown to light red-brown patches, larger and denser towards the blunt pole
Size of eggs:	15×12 mm
Incubation:	13–14 days, beg. from complete clutch
Fledging period:	Nidicolous; remaining 14 days in nest

feeding young, Bavaria, 15.7.1978 (Li)

Nestlings, Amperaue, Bavaria, 19.7.1978 (Li)

Bavaria, 12.5.1977 (Pf)

) Ad., Amperaue, Bavaria, 15.7.1978 (Li)

Willow Warbler (Phylloscopus trochilus)

The Willow Warbler is the commonest Warbler in much of northern Europe. In Finland the population was estimated at 5.7 million pairs. It occurs in light deciduous woodland with a secondary growth of low bushes and tall vegetation, riverine thickets of willow and birch, trees in parkland and roadsides and in the north where there is mixed forest and bush tundra. It rarely occurs in mountains above 1,300 m. It feeds on small insects and larvae, mosquitoes, aphids and small beetles and in autumn on some berries such as elder and currant. It is an active bird, searching for food among the leaves of trees and bushes, often descending to ground-level to feed along ditches and low growth.

The nest is a domed structure of moss, grass, plant stems and roots with a lining of fine plant material and feathers. It is built by the ♀ and is usually placed on the ground in tall vegetation or under the cover of a bush or shrub, often in tree-roots or on a roadside bank. Incubation is by the ♀ alone, leaving the nest for short periods to feed. Both parents feed the young.

The Willow Warbler is entirely migratory, wintering over a large part of tropical Africa south of about 10°N. On passage and in winter it is often seen in loose congregations.

Migration: Departs breeding areas from late July, mostly August–September, returning in April and May.

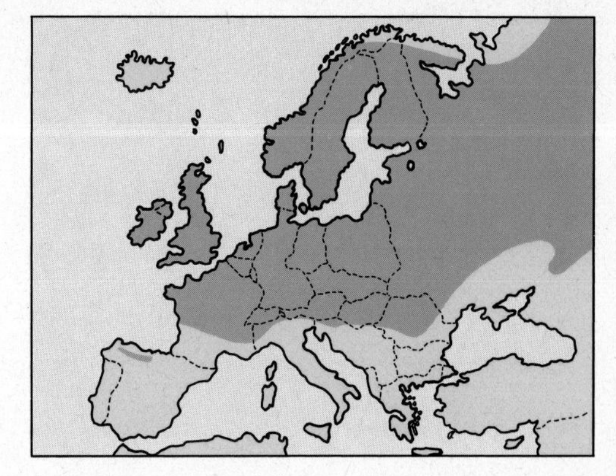

Length:	11 cm
Wing length:	6–7 cm
Weight:	9 g
Voice:	Song: thin and trilling 'teeteeweetweetoo'. Call: 'hoo-eet', similar to Chiffchaff's
Breeding period:	Beginning of May, June. 1 brood per year in the north, more rarely 2 in the south. Replacement clutch possible
Size of clutch:	6–7 (4–8) eggs
Colour of eggs:	Whitish, with fine and rather even light red-brown to rusty spots
Size of eggs:	15.2×12 mm
Incubation:	12–14 days, beg. from complete clutch
Fledging period:	Nidicolous; remaining 15 days in nest

nwald, West Germany, 8.6.1961 (Pf)

Fledged young bird, Bavaria, 22.6.1979 (Li

Bavaria, 12.6.1976 (Pf)

Ad. at nest, Bavaria, 12.6.1979 (Pf)

Wood Warbler (Phylloscopus sibilatrix)

The Wood Warbler is the largest of the European leaf-warblers. It inhabits well-grown woodland, especially beech and oak as well as low oak and birch woods in hilly areas and mixed woodland with fir or spruce. It prefers woods with little secondary growth. It feeds almost entirely on insects at every stage from egg to imago, and midges, beetles and Lepidoptera are also taken. In autumn a few berries may supplement the diet. It feeds mainly in the foliage of trees, but frequents lower growth on migration. It often makes flycatcher-like sallies to catch flying insects.
The nest is a domed structure in a hollow in the ground, among roots or hidden in low vegetation. It is built of dead grass, plant stems, bracken and leaves and lined with fine grass and hair. Nest-building and incubation are by the ♀ alone, though both sexes feed the young.
The Willow Warbler is entirely migratory and winters in central Africa where it frequents evergreen forest.
Migration: Departs breeding areas from late July to August with passage continuing into September. It returns in April and May.

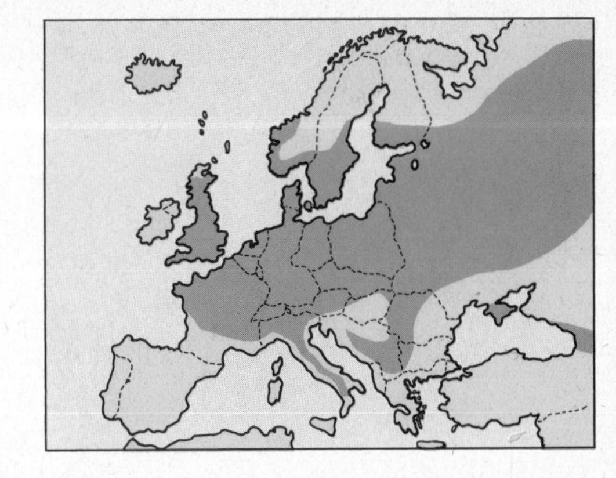

Length:	12.5 cm
Wing length:	6.9–8.0 cm
Weight:	10 g
Voice:	Song: Trilling series of notes with whirring final trill: 'sibsibsibsreel'; also 'pyoo' and 'hwit'
Breeding period:	May, June. 1, rarely 2 broods per year. Replacement clutch possible
Size of clutch:	6–7 (4–8) eggs
Colour of eggs:	White, with numerous red-brown and grey patches, often grouped at the blunt pole, but not in a zone
Size of eggs:	16×12.5 mm
Incubation:	13 days, beg. from complete clutch
Fledging period:	Nidicolous; leaving the nest at 12 days, but tended by the parents a further 4 weeks

., Amperaue, Bavaria, 23.7.1979 (Li)

Nestlings, 7 days old, Bavaria, 19.6.1977 (Pf)

Bavaria, 27.6.1979 (Pf)

t) Ad., Bavaria, 2.6.1976 (Pf)

Arctic Warbler (Phylloscopus borealis)

The Arctic Warbler inhabits sub-arctic birch woods with a good undergrowth of bilberries and similar vegetation. It is also found in the dense taiga forests, usually along rivers or forest edges where there is some birch, alder or willow thickets. It feeds entirely on insects at all stages of development while mosquitoes make up much of the diet. It is active and lively, making flycatcher-like sallies after insects and hovering to pick insects from leaves. The nest is a domed structure placed on the ground among tree-roots, moss or tall herbage. The nest is made of moss grasses and leaves with a lining of fine grass or hair. The ♀ builds the nest and incubates alone, though both sexes rear the young.
The Arctic Warbler is entirely migratory, wintering in tropical Asia. Some birds occur as vagrants in western Europe.
Migration: Departs breeding areas in late July to September, returning in May–June.

Bonelli's Warbler (Phylloscopus bonelli) The Bonelli's Warbler occurs in low oak and beech woods in hilly areas or mountain slopes. It also inhabits mixed beech and fir woods and dry pines. Throughout most of the range it breeds between 600 m and 1,800 m. It feeds entirely on small insects. The nest is usually built into a hollow in the ground on a bank or in undergrowth. It is a domed structure built of grass, roots and similar material, lined with hair. Nest-building and incubation are by ♀ alone.
The Bonelli's Warbler is migratory, wintering in the acacia steppes across Africa south of the Sahara.
Migration: Departs breeding areas in late July to September, returning in April to early May.

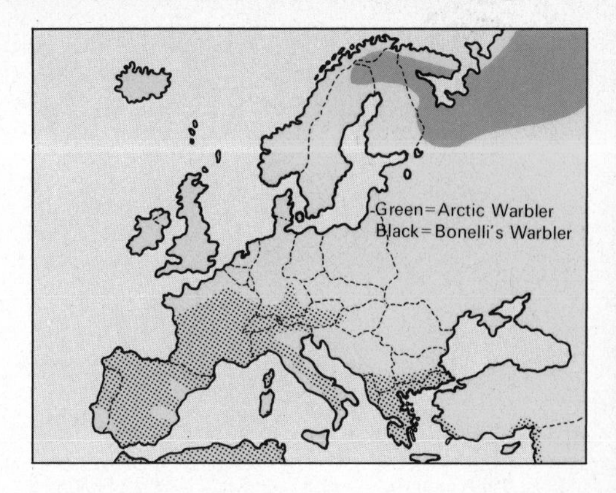

Length:	12 cm — Arctic Warbler
Wing length:	6–7 cm
Weight:	10 g
Voice:	Song is a whirring trill. Call: 'tsik' or 'tsrrrt'
Breeding period:	End of June onwards. 1 brood per year
Size of clutch:	6 (5–7) eggs
Colour of eggs:	Whitish, with sparse but large dark red-brown spots
Size of eggs:	16.2×12.2 mm
Incubation:	13 days, beg. from complete clutch
Fledging period:	Nidicolous; remaining 12 days in nest

Length:	11.5 cm — Bonelli's Warbler
Wing length:	5.8–6.6 cm
Weight:	7 g
Voice:	Song: a trill at one pitch: 'dididididi'. Call: 'hro-eet' and 'chee-chee'
Breeding period:	In the north, late May and June; in the south from mid-May. 1 brood per year
Size of clutch:	5–6 (7) eggs
Colour of eggs:	Whitish with dense fine red-brown spots, concentrated at the blunt pole
Size of eggs:	16×13.1 mm
Incubation:	13 days, beg. from penultimate egg
Fledging period:	Nidicolous; remaining 10–12 days in nest

ctic Warbler, Sweden (Sw)

t) Arctic Warbler, Sweden (Sw)

Chiffchaff (identified as Bonelli's Warbler), Mallorca, 26.3.1976 (Tö)

Goldcrest (Regulus regulus)

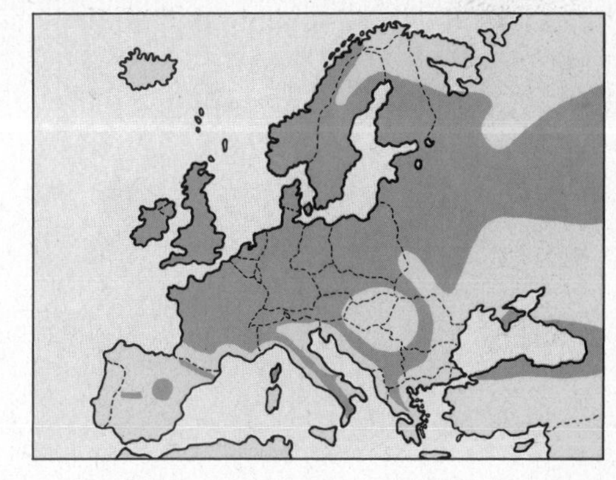

The Goldcrest is found in coniferous forest and in conifers in parks, gardens and mixed woodland. Yew, spruce, fir and cedar are its preferred trees. Outside the breeding season, particularly on migration, it occurs in a variety of habitat including hedgerows and thickets.
It feeds on small insects and spiders and is a very active bird, constantly flitting amongst branches in search of food. It often hangs upside-down in tit fashion. It is very trusting and takes little notice of humans. Outside the breeding season it is often seen in small groups or with mixed feeding-flocks of tits and other birds.
The nest is a deep well-made cup of moss, lichen and spiders' webs, lined with feathers. It is sited in a conifer and is suspended from a fork or twigs near the end of a branch or tucked against a forked branch. Both sexes may assist with nest-building but it is mainly the responsibility of the ♀. She incubates alone but both sexes rear the young.
The Goldcrest is partially migratory. The northern and eastern European populations are entirely migratory, wintering in west and southern Europe. In other areas the species is mainly resident, though it wanders throughout the winter.
Migration: Autumn movements start in August but mainly September–October with passage into November. Return passage takes place in March–April.

Length:	9 cm
Wing length:	5.0–5.5 cm
Weight:	5 g
Voice:	A repeated shrill: 'seeseesee' and 'seeseeseehseretet'. The note is very high and thin
Breeding period:	End of April, May, end of June. 2 broods per year. Replacement clutch possible
Size of clutch:	7–10 (6–13) eggs
Colour of eggs:	Yellowish white with very fine brown spots, so dense as to give the impression of a uniformly pale red-brown egg
Size of eggs:	14×10.2 mm
Incubation:	*ca* 16 days, beg. from complete clutch
Fledging period:	Nidicolous; remaining 15–20 days in nest

♂, Bavaria, 16.10.1972 (Pf)

♀ at nest, Amrum Island, 18.6.1973 (Qu)

Bavaria, 27.5.1973 (Li)

Ad. ♀, Bavaria, 12.9.1972 (Pf)

Firecrest (Regulus ignicapillus)

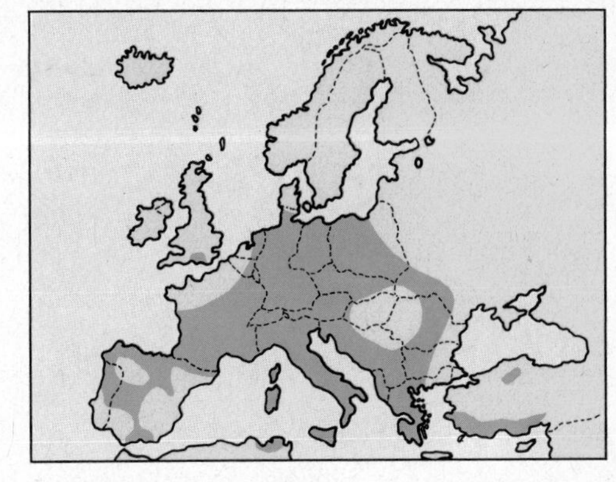

The Firecrest has a more southerly distribution and in general a different habitat-preference to that of the Goldcrest. It is found in broad-leaved and mixed woodland, evergreen oak forest and in fir and pine woods in some areas.

Like the Goldcrest, it feeds entirely on small insects at all stages of development, including spiders. Its habits closely resemble those of the Goldcrest though it is more often seen in low cover and is less gregarious, rarely being seen in small flocks.

The nest is rather neater and smaller than that of the Goldcrest and in a similar site, usually an evergreen, conifer or in ivy. Incubation is by the ♀ alone though both parents feed the young. The Firecrest is largely resident and sedentary and, though it wanders in winter, it rarely associated with tits and other birds. In the north and east of the range it is migratory, wintering in western and southern Europe. It occurs as a passage migrant in Britain, particularly in autumn, and from the early 1960s it has nested regularly. It is now quite widespread throughout southern England. This colonisation seems to be part of a northward extension to the range, as it has increased in numbers in the north of its continental range and bred in Denmark for the first time in 1961.

Migration: Autumn dispersal from September but mainly October to early November, returning in March–April.

Length:	9 cm
Wing length:	5.0–5.5 cm
Weight:	5 g
Voice:	'Sisisisiiss', with increasing volume; lower pitched and stronger than Goldcrest
Breeding period:	Early May, June, July. 1–2 broods per year
Size of clutch:	7–12 eggs
Colour of eggs:	Noticeably more reddish or salmon-coloured than Goldcrest, with very fine spots
Size of eggs:	13.5×10.6 mm
Incubation:	14–15 days, beg. from last egg
Fledging period:	Nidicolous; fledged at 19–20 days

Bavaria, 18.6.1979 (Li)

Fledged young bird, Bavaria, 26.6.1979 (Li)

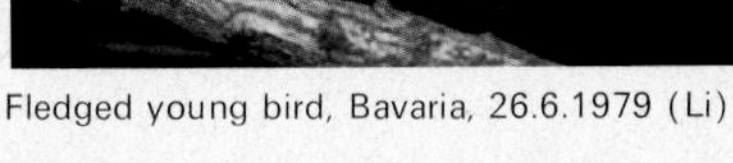

Bavaria, 22.5.1979 (Li)

) ♀ at nest, Bavaria, 18.6.1979 (Li)

Spotted Flycatcher (Muscicapa striata)

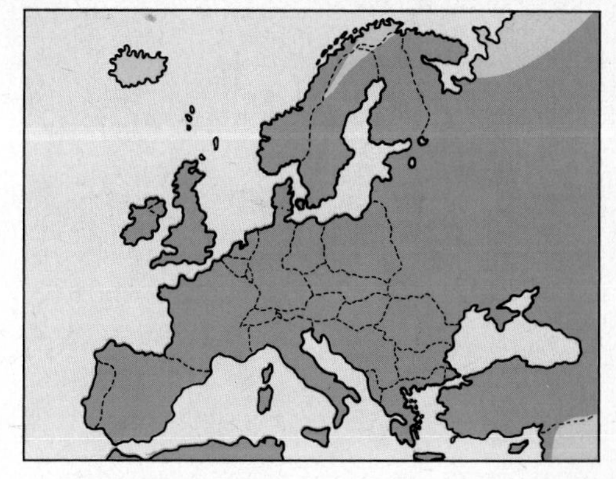

The Spotted Flycatcher is widely distributed throughout Europe and occurs in a variety of different habitats. It is found in open woodland, mainly deciduous but conifers in some areas, parkland, gardens, hedgerows with tall trees, roadsides with lines of trees and in the north of the range into sub-arctic birch woods.
It feeds almost exclusively on insects, mostly caught on the wing. It will also take earthworms, though it is rarely seen on the ground, and it has been recorded taking berries in autumn. It usually perches on low branches of trees or on fences and makes brief sallies to catch insects, often returning to the same perch. It is restless and often flicks its wings and tail, though less so than the other flycatchers.
The nest-sites are varied, It may use natural holes in trees, man-made holes such as pipes, holes in walls, nest-boxes, etc. It also nests in creepers, behind dead bark or against the trunk of a tree. The nest is a cup of grasses, roots, lichen, twigs and plant down, lined with fine plant materials, feathers, hair or dead leaves. Both sexes build the nest but this is mainly done by the ♀. Incubation is usually by the ♀ alone. At first the ♀ feeds the young with food brought by the ♂, but later both parents feed them.
The Spotted Flycatcher is entirely migratory, wintering in much of Africa south of about 10°N.
Migration: Autumn movements from late July to October, mainly August–September. Return passage in April to early June.

Length:	14 cm
Wing length:	8.1–8.9 cm
Weight:	19 g
Voice:	Shrill 'tzee' and 'tzek-tzuk'. Song: a series of squeaky notes, rather infrequent
Breeding period:	End of May, June. 1–2 broods per year. Replacement clutch possible
Size of clutch:	4–5 (6) eggs
Colour of eggs:	Whitish green, with variable coarse red-brown spots, concentrated at the blunt or sharp pole
Size of eggs:	18.4×14.1 mm
Incubation:	11–15 days, beg. from complete clutch
Fledging period:	Nidicolous; remaining *ca* 2 weeks in nest, fed by adults for further 2–3 weeks

sfer of food to brooding ♀, Bavaria, 1972 (Li)

Young shortly before fledging, Bavaria, 1972 (Li)

Bavaria, 8.6.1978 (Pf)

) Amperaue, Bavaria, 1972 (Li)

Pied Flycatcher (Ficedula hypoleuca)

The Pied Flycatcher breeds in deciduous and coniferous woodlands, open woodland, cultivated areas with scattered trees, parkland and gardens. In some areas it is found in upland oak-woods and it prefers mature woodland with well-spaced trees. Much of the original habitat has disappeared from western Europe. It takes to nest-boxes readily and provision of these has increased the population in some areas. It feeds almost entirely upon insects and their larvae. Some worms are taken, especially in the north of the range, and it sometimes takes berries in the autumn. It is essentially an active, restless bird, flicking its wings and tail a great deal. After flycatching sallies it rarely returns to the same perch. It keeps mostly to the canopy of trees, though will feed in lower growth and sometimes on the ground.
The nest-site is in a hole in a tree, often an abandoned woodpecker's hole, a crevice in a wall or building or in a nest-box. The nest is a loosely-made cup of grass, leaves, roots and moss, lined with hair or fine plant material, sometimes feathers. It is built by the ♀, who also incubates alone, being fed by the ♂ whilst brooding. The young are fed mainly by the ♀, though the ♂ assists at times.
The Pied Flycatcher is entirely migratory, wintering in tropical west and central Africa.
Migration: Departs breeding areas in August–October, returning in April–June.

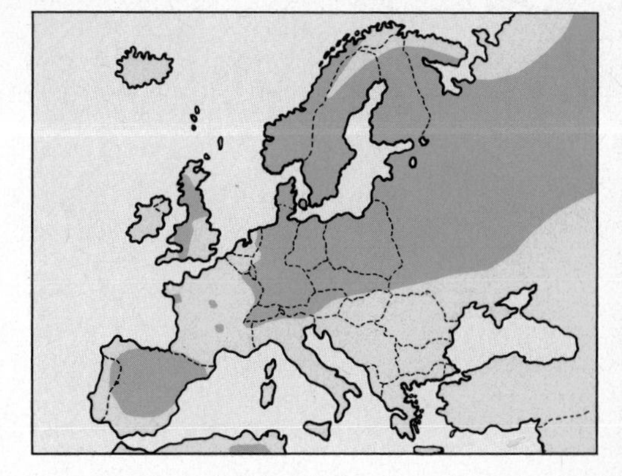

Length:	13 cm
Wing length:	7.5–8.2 cm
Weight:	13 g
Voice:	Snapping 'check', and 'whitt'. Song: 'zee-it zee-it zee-it' with Redstart-like trilling
Breeding period:	May, June. 1 brood per year. Replacement clutch possible
Size of clutch:	6–7 (4–9) eggs
Colour of eggs:	Pale blue, without markings
Size of eggs:	17.9×13.4 mm
Incubation:	12–14 days, beg. from complete clutch
Fledging period:	Nidicolous; leaving nesting-hole at 14 days

f browner central European form, Denmark (Gé)

) Ad. ♂, Smaland, Sweden, 24.6.1963 (Chr)

♀ at nesting-hole, North Rhine-Westphalia, May 1975 (Sieb)

Baden-Württemberg, West Germany, 1975 (Schw)

Collared Flycatcher (Ficedula albicollis)

The Collared Flycatcher inhabits old beech- and oak-woods, parks, gardens and orchards with mature trees. It does not occur in coniferous woodland or the lighter woodlands inhabited by the Pied Flycatcher. It also breeds at a higher altitude than that species, mostly in montane beech-woods. In some areas it has been recorded as interbreeding with the Pied Flycatcher.

It feeds almost exclusively on insects and larvae and perhaps berries in autumn. It is most frequently seen in the tree-tops and rarely descends to lower vegetation, unlike the Pied Flycatcher. Otherwise its habits and behaviour closely resemble that species.

The nest is much as the Pied Flycatcher's and in a similar site. The Collared Flycatcher is migratory, wintering in central and East Africa, where its range does not overlap with that of the Pied Flycatcher.

An eastern form, the **Semi-collared Flycatcher** (*Ficedula semitorquata*), is now treated as a separate species. It occurs in Greece and Asia Minor and probably has habits and breeding biology very similar to those of the Collared Flycatcher.

Migration: Departs breeding areas in August–September, returning in April to early June.

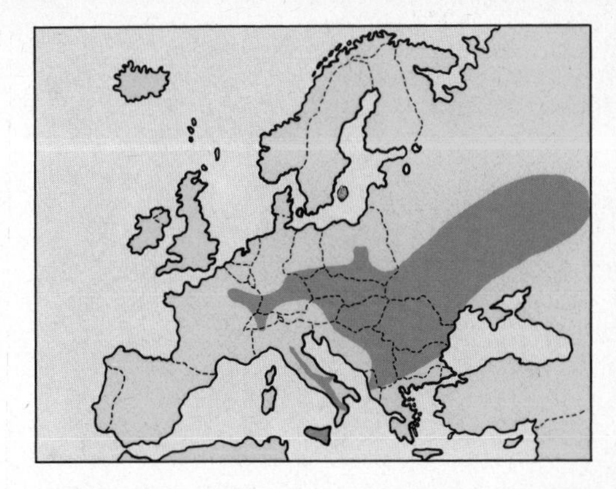

Length:	13 cm
Wing length:	7.8–8.5 cm
Weight:	10 g
Voice:	Calls: like Pied Flycatcher. Song: 'sit-sit-sit-siu-see'
Breeding period:	May, early June. 1 brood per year. Replacement clutch possible
Size of clutch:	5–7 (–9) eggs
Colour of eggs:	Light blue, without markings
Size of eggs:	17×13.4 mm
Incubation:	12–15 days, beg. from complete clutch
Fledging period:	Nidicolous; leaving breeding-hole at 16 days

ith nesting material, Bavaria, June 1976 (Pf)

♀, Bavaria, June 1976 (Pf)

Bavaria, 18.5.1976 (Pf)

t) ♂ in front of nesting-hole, Bavaria, June 1976 (Pf)

Red-breasted Flycatcher

(Ficedula parva)

The Red-breasted Flycatcher inhabits mature deciduous woodland, particularly beech and hornbeam, as well as mixed and coniferous forest, the latter mainly in montane areas. It is fond of sites near water and along paths or trails in woodland.
The diet is chiefly insects, mainly caught on the wing. It also takes some larvae and perhaps a few berries in the autumn. It is a rather shy and secretive bird, keeping to the canopy of the woodland, though, being active and restless, it is often seen on flycatching sallies from the trees. It will perch on fences and sometimes take food from the ground.
The nest-site is usually in a hollow or crack in a tree-trunk, sometimes in a hole in a tree or wall or supported by twigs next to the trunk of a tree. It is a small cup of moss with dead leaves, spiders' webs, and lichen, lined with hair. It is built mostly by the ♀ who also incubates alone, being fed on the nest by the ♂. Both parents tend the young.
The Red-breasted Flycatcher is entirely migratory, wintering in tropical Asia. Perhaps a few winter in Africa as it is regularly on passage in parts of North Africa. Like the Barred Warbler, it has a tendency to reverse the south-east orientation in autumn and, as a result, occurs well to the north-west of its breeding range. It has gradually extended its breeding range to the west in recent times.
Migration: Departs breeding areas in August–October, returning in late April–May.

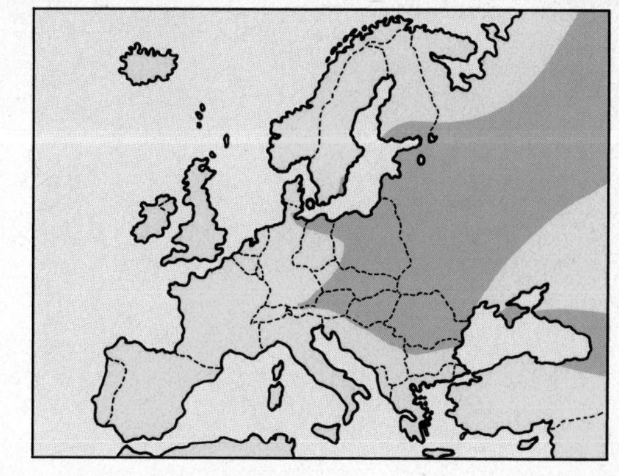

Length:	11.5 cm
Wing length:	6.4–7.1 cm
Weight:	10 g
Voice:	Calls: a sharp 'chick', also 'pfiff' and a wren-like chatter. Song resembles Wood or Willow Warbler
Breeding period:	Beginning of May, June. 1 brood per year. Replacement clutch possible
Size of clutch:	5–6 (4–7) eggs
Colour of eggs:	Greenish to yellowish white with many fine red-brown spots, often in a crown round the blunt pole
Size of eggs:	16×13 mm
Incubation:	12 days, beg. from complete clutch
Fledging period:	Nidicolous; leaving nest at 2 weeks

nest, Austria, 25.6.1969 (Fe)

Young birds, Austria, 28.6.1969 (Fe)

Austria, 7.6.1960 (Fe)

♂ at nest, Austria, 25.6.1969 (Fe)

Siberian Rubythroat (Luscinia calliope)

The Siberian Rubythroat lives in the taiga forest of Siberia, and occurs as a breeding bird in the extreme north-east of Europe. It is found in dark, swampy forests with thick undergrowth and in mixed forest with decaying trees. It feeds on insects and larvae, small worms and other invertebrates, and some berries are taken in the autumn. It is a secretive bird. The nest is on the ground, in a tussock or in vegetation under a bush. It is a loose cup of plant stems, grass and plant material, lined with hair or plant down. Incubation and care of the young are by the ♀ alone.

It is entirely migratory, wintering in tropical Asia. It is a rare vagrant to western Europe.

Migration: Departs breeding areas in September–October, returning in late April to early June.

Red-flanked Bluetail (Tarsiger cyanurus) The Red-flanked Bluetail inhabits taiga forest as well as birch-woods and subalpine coniferous forest. It is mainly insectivorous and catches its food in a Redstart-like manner. It prefers to perch in the lower branches of trees. The nest is sited in a hollow log, hole in a bank or among tree-roots. It is a cup of grasses, moss and roots lined with fine plant material, hair or pine needles. Incubation is by the ♀ alone.

It is migratory, wintering mainly in tropical south-east Asia. It is a vagrant in western Europe.

Migration: Departs breeding areas in September–October, returning in late April to May.

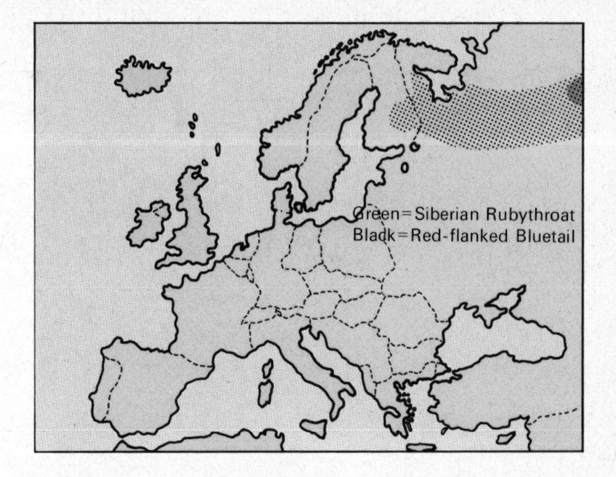

Length:	14–15.5 cm Siberian Rubythroat
Wing length:	6.7–8.8 cm
Weight:	19.5–23 g
Voice:	Call: 'tiuit-tiuit'. Song: loud, Nightingale-like, often imitative
Breeding period:	End of May–June. 1 brood per year
Size of clutch:	5 (4–6) eggs
Colour of eggs:	Greenish blue, with sparse fine red-brown spots
Size of eggs:	21×15.5 mm
Incubation:	14 days, beg. from complete clutch
Fledging period:	Nidicolous; leaving nest at 12–13 days

Length:	14 cm Red-flanked Bluetail
Wing length:	7–8 cm
Weight:	14–15 g
Voice:	Call: Robin-like 'tic tic'. Song: soft and thrush-like
Breeding period:	Mid-May, June. 1 brood per year
Size of clutch:	3–6 eggs
Colour of eggs:	Whitish, with sparse fine rusty spots, often denser at the blunt pole
Size of eggs:	17.8×14 mm
Incubation:	Unknown
Fledging period:	Nidicolous; remaining *ca* 14 days in nest

-flanked Bluetail imm., ♂, Bavaria, 31.10.1971 (Wü)

Red-flanked Bluetail imm., ♂, Bavaria, 31.10.1971 (Wü)

t) Siberian Rubythroat, ad. ♂, Voliere (Pf)

Nightingale (Luscinia megarhynchos)

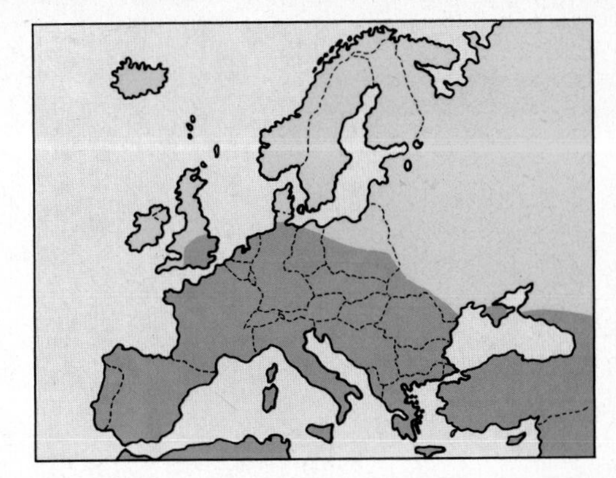

The Nightingale inhabits broad-leaved woodland and copses with a thick understorey of brambles, nettles, wild rose and other plants. It also occurs in thickets and hedgerows in open country, gardens and parks where there are dark shady places under thick cover.
It feeds on insects, especially beetles, also worms, spiders and berries. Most of the food is obtained on the ground, usually under the cover of foliage. It is a secretive bird, keeping to thick cover, dropping down to pick up insects from the ground. The beautiful song is usually given from low cover, but sometimes it will sing when perched in an exposed place. It sings throughout the day and habitually by night.
The nest is on the ground or in lush vegetation in shrubby growth. It is a loose cup of grass, leaves and twigs lined with fine grass, roots or hair. The ♀ is responsible for nest-building and incubation, but both parents feed the young.
The Nightingale is entirely migratory, wintering in tropical Africa. It has decreased in numbers in the north of its breeding range, probably due to destruction of habitat with increased agricultural development. It is still very common in southern Europe.
Migration: Departs breeding areas in late July to September and returns in April and May.

Length:	17 cm
Wing length:	7.8–8.9 cm
Weight:	22 g
Voice:	Calls: 'tacc-tacc' and 'hooet'. Song: 'Chooc-chooc-chooc-chooc dyu-dyu-dyu', fluty and ending in crescendo
Breeding period:	May to mid-June. 1 brood per year. Frequent replacement clutch
Size of clutch:	5 (4–6) eggs
Colour of eggs:	Greenish white to blue-green, with dense olive-brown or olive-green spots, giving monochrome effect
Size of eggs:	20.8 × 15.6 mm
Incubation:	14 days, beg. from complete clutch
Fledging period:	Nidicolous; leaving the nest at 11–12 days, unable to fly; independent at 4 weeks

ged young bird, Greece, 19.6.1975 (Pf)

Nestlings, Greece, 12.6.1975 (Pf)

Anatolia, Turkey, 15.5.1975 (Li)

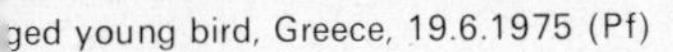

Ad. at nest (1½ m above ground), Greece, 12.6.1975 (Pf)

Thrush Nightingale or Sprosser

(Luscinia luscinia)

The Thrush Nightingale is an eastern sibling of the Nightingale. It is found in open deciduous woodland with a low groundcover of nettles, bramble, etc, as well as in damp alder and birch thickets and swampy places with dense cover. It does not occur in parks and gardens as frequently as the Nightingale and prefers damp areas, often along streams or rivers.

The diet is mainly small worms and a variety of insects, berries of elder and currants, etc. are taken in the autumn. Like the Nightingale, it takes most of its food from the ground. It is a secretive bird, keeping to dense cover, and often runs rather than takes flight when disturbed.

The nest-site is usually a hollow in the ground among vegetation or roots and fallen branches, etc. The nest is a cup of dead leaves, plant stems, grass and small twigs, lined with fine plant matter, roots and hair. It is built by the ♀, who also incubates alone, being fed by the ♂ whilst brooding. Both parents tend the young.

The Thrush Nightingale is a migrant, wintering in tropical East Africa. It has extended its European range to the north in recent years and, though still a vagrant to Britain, it has become more regular.

Migration: Departs breeding areas from late July to September, returning in late April to May and early June in the north.

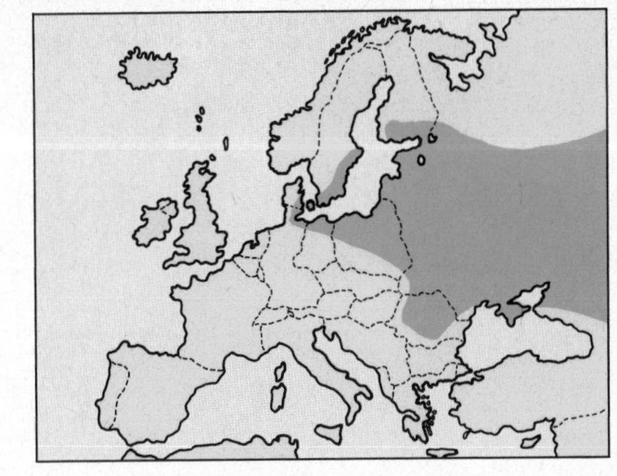

Length:	17 cm
Wing length:	8.4–9.2 cm
Weight:	22 g
Voice:	Very similar to Nightingale but without crescendo. Call: 'whit'
Breeding period:	May, June. 1 brood per year. Replacement clutch possible
Size of clutch:	4–5 (6) eggs
Colour of eggs:	Olive-brown or -green, monochrome effect; not distinguishable from Nightingale's
Size of eggs:	21.7×16.4 mm
Incubation:	13 days, beg. from complete clutch
Fledging period:	Nidicolous; leaving nest at 11 days, unable to fly; require a further 10 days till able to fly

reservation photograph (Li)

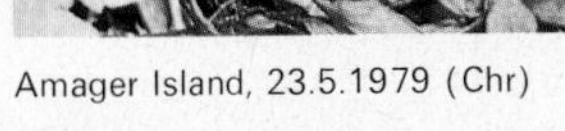

Amager Island, 23.5.1979 (Chr)

) Ad. at nest, Amager Island, Denmark, June 1979 (Chr)

Bluethroat (Luscinia svecica)

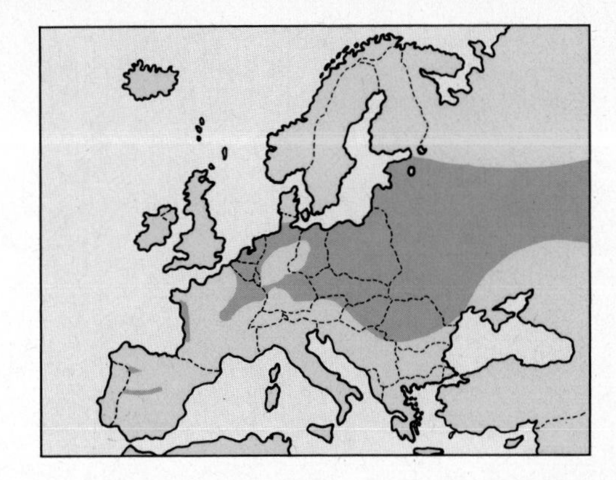

Two races of the Bluethroat breed in Europe. The White-spotted form (*L. s. cyanecula*) occurs in bushy areas in damp surroundings, scrubby growth of willows, alder with reeds and other rank vegetation along streams and rivers as well as mountain slopes with broom and other plants. On passage it is found in cultivated areas, hedgerows, gardens and bushy areas.
It feeds on small insects, including beetles and aquatic insects and larvae as well as worms, and in autumn on small berries. It feeds mostly on the ground, usually under dense cover but will come out into the open if undisturbed. It is generally very secretive and skulking, moving about in cover with great agility, running mouse-like on the ground. In winter quarters it will perch on exposed posts, wires and plants like a Stonechat.
The nest-site is on the ground, usually very well concealed in dense vegetation on the side of a small bank. The nest is a cup of plant stems, grass, moss and roots lined with fine plant matter, hair and sometimes feathers. Nest-building and incubation are chiefly the ♀'s responsibility, though the ♂ may assist at times. Both sexes feed the young.
The White-spotted Bluethroat is a migrant, wintering in Africa in a belt south of the Sahara as well as in North Africa and extreme south-west Europe.
Migration: Departs breeding areas from late July to September, returning in March–April.

Length:	14 cm
Wing length:	7.1–8.1 cm
Weight:	18 g
Voice:	'Wheet' and 'tac'. Song: loud and varied, recalling Nightingale or Woodlark
Breeding period:	End of April, May, June. 1, rarely 2 broods per year. Replacement clutch possible
Size of clutch:	5–6 (–9) eggs
Colour of eggs:	Grey-green to blue-green, with dense fine red-brown spots
Size of eggs:	19×14 mm
Incubation:	14 days, beg. from complete clutch
Fledging period:	Nidicolous; leaving nest at 2 weeks, not yet able to fly

varia, 31.5.1970 (Li)

Nestlings, Bavaria, 18.5.1974 (Pf)

Bavaria, 15.5.1970 (Li)

Ad. ♂, Bavaria, 12.5.1972 (Pf)

Red-spotted Bluethroat

(Luscinia svecica svecica)

The Red-spotted form of the Bluethroat (*L. sv. svecica*) has a more northerly distribution than the White-spotted form. It inhabits willow and birch thickets in shrub tundra, mainly in swampy places, open marshland with scattered willows and other damp areas. In winter if is found in open areas including crops in agricultural land, reed-beds, and vegetation along the banks of rivers and streams.
The diet is much the same as for the White-spotted form, though it probably takes more berries in the tundra regions. The habits, breeding biology and nest-site are much as for the White-spotted Bluethroat, though the nests tend to be in more open sites with lower vegetation.
It winters in Africa south of the Sahara, North Africa and the Middle East and in extreme south-west Europe. On passage it is a regular visitor to Britain, though often skulking and difficult to see.
Migration: Departs breeding areas in August–September, passage continuing into October and, exceptionally, November. Return passage in April to early June.

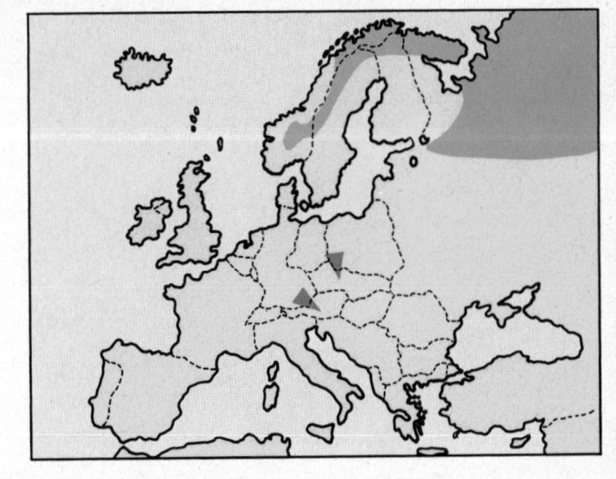

Length:	14 cm
Wing length:	♂ 6.9–7.5 cm; ♀ 6.5–7.4 cm
Weight:	17–22 g
Voice:	Like *Luscinia sv. cyanecula*
Breeding period:	Mid-June, early July. 1 brood per year. Replacement clutch possible
Size of clutch:	6–7 (5–8) eggs
Colour of eggs:	Blue-green, with dense fine rufous spots, often concentrated at the blunt pole
Size of eggs:	18.6×14.1 mm
Incubation:	13–14 days
Fledging period:	Nidicolous; at 14 days, unable yet to fly, they hide in the nearby vegetation

in transitional plumage (Pf)

♀ carrying food, June 1975 (Pf)

Fledged young bird, Dovrefjell, Norway, 12.8.1980 (Pf)

eft) Ad. ♂, Fokstua, Norway, 25.6.1974 (Pf)

Robin (Erithacus rubecula)

In Britain the Robin is a familiar garden bird, often tame and trusting. This is not the case with continental Robins, which are much shyer and avoid living in close contact with man. The habitats therefore differ, as does the nest-site. The Robin is found in thick broad-leaved woodland with a well-developed undergrowth as well as in coniferous woodland, including dark spruce forests with many dead trees, fallen trunks and mossy cover. In Britain it is also found in parks, gardens and agricultural land.
It feeds mainly on insects and their larvae, spiders, earthworms, seeds, soft fruit and berries. It feeds mainly on the ground, often in the open, particularly in Britain, and in the cover of thick vegetation. It is quite pugnacious towards other Robins and will chase intruders from the territory.
The nest-site is in a hole in a hollow tree-stump, in a bank or among roots. In Britain a hole in a wall is frequently used. The nest is a large cup of dead grass, leaves and moss, lined with hair and fine plant materials, sometimes feathers. It is built by the ♀ and she also incubates alone, being fed by the ♂. Both parents feed the young, though at first the ♂ provides all the food.
Robins are mainly resident and sedentary in western Europe, but the northern and eastern populations move to spend the winter in southern Europe, North Africa and parts of the Middle East.
Migration: Autumn movements in September–October, returning in March–April.

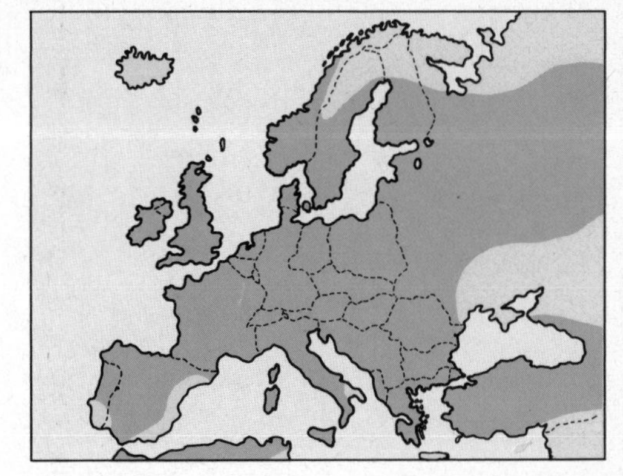

Length:	14 cm
Wing length:	6.9–7.6 cm
Weight:	16 g
Voice:	Song: molodious warble, rather thin and shrill. Calls: 'tseeh' and 'tic' often repeated
Breeding period:	Late March to end of July. 2, sometimes 3 broods per year. Replacement clutch possible
Size of clutch:	5–7 (4–9) eggs
Colour of eggs:	White, with dense fine rufous spots, divided or in a zone at the blunt pole
Size of eggs:	19.8 × 15.4 mm
Incubation:	14 days, beg. from complete clutch
Fledging period:	Nidicolous; remaining 2 weeks in nest

., Bavaria, March 1973 (Pf)

Young birds, Bavaria, May 1974 (Pf)

Bavaria, May 1979 (Pf)

t) Bavaria, June 1970 (Li)

Redstart (Phoenicurus phoenicurus)

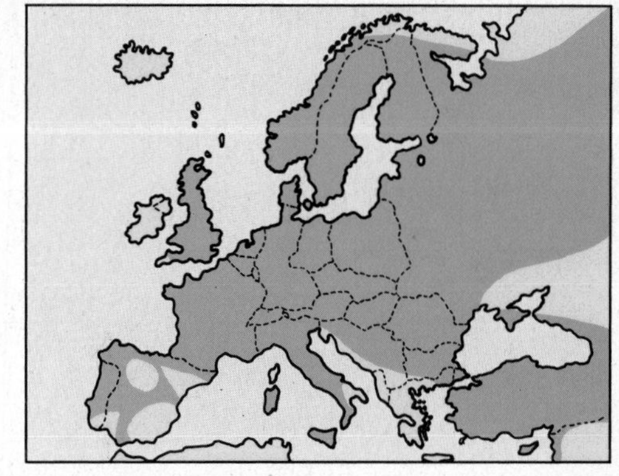

The Redstart inhabits open deciduous woodland, particularly with well-spaced trees and an understorey with open patches. It is also found in mixed woodland and even pure pine forests. In other areas it is found in parks, orchards, large gardens and the arctic birch-zone in the north.
It feeds mainly on insects including Lepidoptera and their larvae, beetles, and sawfly larvae, as well as spiders, small worms and some berries in autumn. The young are fed chiefly on caterpillars. It is mainly arboreal, taking food from among the foliage of the canopy as well as fluttering and hovering for insects on the wing.
The nest is usually in a hole in a tree or stump, often in an abandoned woodpeckers' nest. Sometimes the nest is in a hole in a wall or cliff, on the ground among tree-roots or in a hollow on a bank. The old nests of Swallows are occasionally used. The nest is a loose cup of grass, moss roots and plant materials lined with hair and feathers. It is built by the ♀, who also incubates alone. Both sexes feed the young. The species readily takes to nest-boxes and provision of these will increase the population-density in areas where natural sites are scarce.
The Redstart is entirely migratory, wintering in a belt across northern tropical Africa.
Migration: Departs breeding areas in August–October, returning in April–May.

Length:	14 cm
Wing length:	7.1–8.3 cm
Weight:	15 g
Voice:	Calls: 'hoo-eet' and 'twick'. Song: brief melodious jangle, Robin-like warble
Breeding period:	May, June, 1, rarely 2 broods per year. Replacement clutch possible
Size of clutch:	6–7 (5–10) eggs
Colour of eggs:	Uniform bluish green
Size of eggs:	18.4×13.8 mm
Incubation:	*ca* 14 days, beg. from complete clutch
Fledging period:	Nidicolous; remaining 14 days in nest

♀, Württemberg, West Germany (Schu)

Fledged young bird, Bavaria, 17.6.1973 (Sch)

Clutch in nesting-box, Bavaria, May 1976 (Pf)

) Ad. ♂, Finland (Hau)

Black Redstart (Phoenicurus ochrurus)

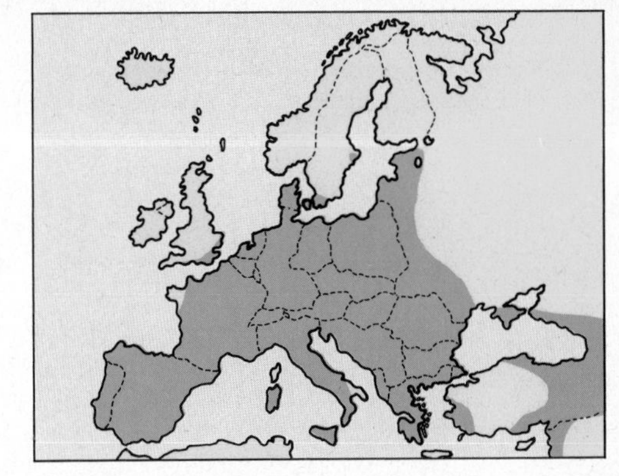

The Black Redstart inhabits rocky areas such as cliffs, mountains, quarries, sea-cliffs and large buildings, derelict building-sites, towers and large industrial buildings. It occurs from sea-level to near the snow-line in high mountains.

It feeds mostly on small insects especially beetles, ants and Lepidoptera as well as spiders, millipedes and some berries. Birds wintering on the coast will take small crustaceans and perhaps molluscs. Though its habits are similar to those of the Redstart, it does not keep to cover but perches in exposed places such as rocks, walls, posts and wires. It feeds mainly on the ground, though it will take insects on the wing and can hover in flight.

The nest-site is usually a crack or crevice in a rock-face, in outcrops or among boulders, as well as in holes in walls and buildings and on ledges. The nest is a rather loose cup made of plant stems, grass, moss and other plant matter, lined with hair, wool and feathers. It is built by the ♀, who also incubates alone. The young are fed by both parents.

Except in the south and west of the range, the Black Redstart is migratory. It winters in western and southern Europe (including parts of Britain), North Africa and the Middle East.

The species has extended its range into northern Europe, breeding regularly in Britain from the 1920s and increasing in the 1940s, when bomb-sites provided a suitable nesting-place. In Britain it winters mainly on sea-coasts.

Migration: Autumn dispersal from August to November, return movements in February–April.

Length:	14 cm
Wing length:	8.0–9.1 cm
Weight:	17 g
Voice:	Call: 'tsip', also -tucc-tucc'. Song: like Redstart's but less rich and softer with remarkable rattling notes
Breeding period:	Beginning of April, May–July. 2 broods per year. Replacement clutch possible
Size of clutch:	5 (4–7) eggs
Colour of eggs:	Pure white
Size of eggs:	20×15 mm
Incubation:	12–16 days, beg. from complete clutch
Fledging period:	Nidicolous; leaving nest at about 2 weeks but not yet able to fly

♀ at nesting-hole, Bavaria, July 1971 (Li)

Fledged young bird, Bavaria, 26.6.1977 (Li)

Bavaria, May 1971 (Li)

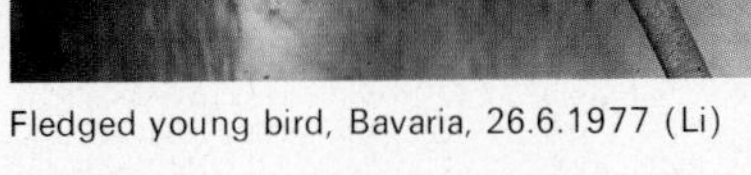

) Ad. ♂, Northern Swabia, West Germany, May 1972 (Par)

Stonechat (Saxicola torquata)

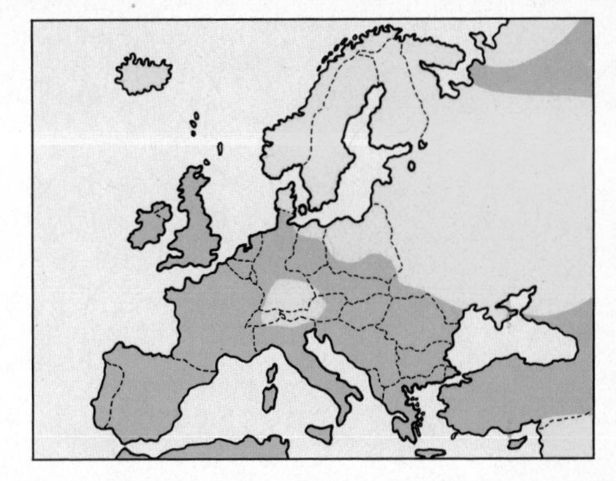

The Stonechat is a bird of open country, heathlands, grassy areas, alpine meadows, scrubby country and agricultural land. It prefers dry, sandy soils with a good cover of bushes and scattered trees. It occurs from sea-level up to about 1,200 m, though much higher outside Europe.
It feeds on a variety of insects including Lepidoptera, beetles, and ants, as well as spiders, worms and a few seeds. It perches freely in exposed places such as the tops of bushes, fence-wire, posts and telegraph-wires. Most of the food is taken from the ground, though it will chase insects on the wing.
The nest is usually sited on the ground at the foot of a bush such as gorse or bramble; sometimes it is in vegetation a little above ground. The nest is a cup of grass, moss and sometimes plant stems, lined with hair, wool or feathers. It is built by the ♀, who also incubates alone, though sometimes the ♂ may assist. Both parents feed the young.
In western and southern Europe the species is mainly resident and may suffer greatly from the effects of a hard winter. The eastern population is migratory. Most winter in southern Europe, North Africa and the Middle East, but some cross the Sahara to winter in tropical Africa. The species has extended its range to the north in recent years, colonising parts of Norway in the 1970s.
Migration: Departs breeding areas in September–October, returning in March–April.

Length:	13 cm
Wing length:	6.4–7 cm
Weight:	14 g
Voice:	Calls: 'tsak task' like two pebbles struck together. Song resembles that of Dunnock
Breeding period:	March–May, June. 2, sometimes 3 broods per year
Size of clutch:	5–6 (–8) eggs
Colour of eggs:	Greenish blue, with fine dense pale rufous spots, often concentrated in a zone at the blunt pole
Size of eggs:	18.9×14.3 mm
Incubation:	14 days, beg. from complete clutch
Fledging period:	Nidicolous; leaving nest, not yet able to fly, at about 2 weeks

♀, Lower Saxony, West Germany, 28.7.1973 (Pf)

Nestlings, Lower Saxony, 6.8.1961 (Sy)

Turkey, 28.6.1977 (Li)

) Ad. ♂, Seewinkel, Austria, 1968 (Li)

Whinchat (Saxicola rubetra)

The Whinchat is found in open grassland with sparse bushes, marshy meadows, hayfields, swampy heathland, moorland, roadside verges, the edges of forests and clearings in woodland. It tolerates more open habitat than the Stonechat and is particularly fond of areas with a growth of umbelliferous plants.
The diet is mainly insects, including beetles, Lepidoptera larvae, some spiders, earthworms and small molluscs. Much of the food is obtained from umbellifers, with flower-visiting insects being prominent.
It perches freely on the tops of bushes and low plants as well as on fences and telephone-wires. It is rather crepuscular in habits, often remaining active till dark.
The nest-site is on the ground, usually in the shelter of a small bush or in tall vegetation, and always well hidden. The nest is a cup of grasses and moss, lined with fine grass and hair. It is built by the ♀ with the ♂ in attendance. Incubation is by the ♀ alone and both parents feed the young.
The Whinchat is a migrant, wintering in tropical Africa over a wide area.
Migration: Departs breeding areas from mid-July to early October, mostly August–September, returning in April and May.

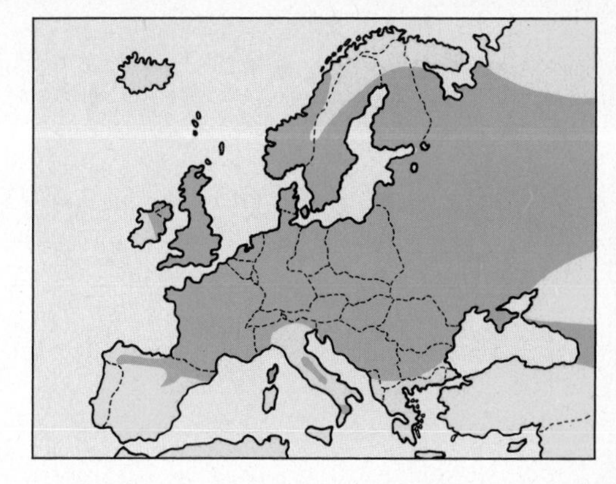

Length:	13 cm
Wing length:	7.3–8.1 cm
Weight:	18 g
Voice:	Call: 'tic tic' and 'tu'. Song: brief metallic warble like Stonechat or Redstart
Breeding period:	Mid-May, June. 1 or 2 broods per year. Replacement clutch possible
Size of clutch:	5–6 (–7) eggs
Colour of eggs:	Brilliant light blue, rarely with fine red-brown speckles
Size of eggs:	19×15 mm
Incubation:	13 days, beg. from complete clutch
Fledging period:	Nidicolous; remaining 12–14 days in nest; then not able to fly, and only independent after a further month

ustria, 1968 (Li)

Imm., Schleswig-Holstein, 10.6.1978 (Bey)

Bavaria, 8.6.1976 (Pf)

) ♂, Austria, 1968 (Li)

Isabelline Wheatear

(Oenanthe isabellina)

The Isabelline Wheatear occurs in the extreme south-east of the region, and is found in desert and semi-desert areas, grassy steppe and plains and bare hillsides. Similar arid habitat is occupied in the winter quarters.
The diet is mainly insects and their larvae and some small seeds. Most of the food is obtained on the ground, the bird often perching on a rock or similar exposed perch and dropping onto food items when spotted. The ♂ has a remarkable acrobatic song-flight, shooting into the air and performing 'stunts' or hovering whilst giving its loud song.
The nest-site is in a hole or crack in the ground, rodent burrows or among low boulders. Sometimes it may excavate or partially excavate its own burrow. The nest is a shallow cup of grass, plant stems, hair and wool built mainly by the ♀. Incubation is by the ♀ alone and both parents feed the young.
The Isabelline Wheatear is a migrant, wintering in a belt across Africa south of the Sahara, as far west as Senegal, which is remarkable in view of its eastern breeding range. A few birds overwinter north of the Sahara in North Africa and the Middle East.
Migration: Departs breeding areas in September–October, returning in March–April.

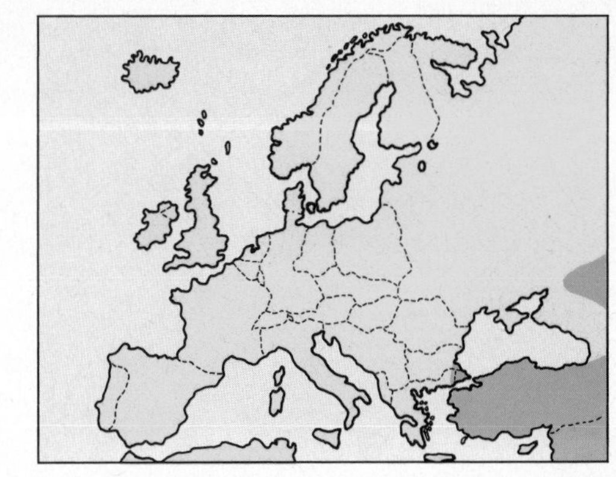

Length:	16.5 cm
Wing length:	9.4–10.7 cm
Weight:	27–31 g
Voice:	Song: loud and Lark-like with astonishing imitation of other birds. Call: 'wheet-whit' and 'cheep'
Breeding period:	Late March, May–June. 2 broods per year
Size of clutch:	5–6 (4–7) eggs
Colour of eggs:	Monochrome pale blue, rarely with a few fine reddish spots
Size of eggs:	22.8 × 17 mm
Incubation:	14 days, beg. from complete clutch
Fledging period:	Nidicolous; leaves nest when still unable to fly but keeps to surrounding cover of stones and burrows

Nestlings, Turkey, 12.6.1973 (Li)

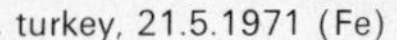

, turkey, 21.5.1971 (Fe)

?) Ad., Turkey, 12.6.1973 (Li)

Wheatear (Oenanthe oenanthe)

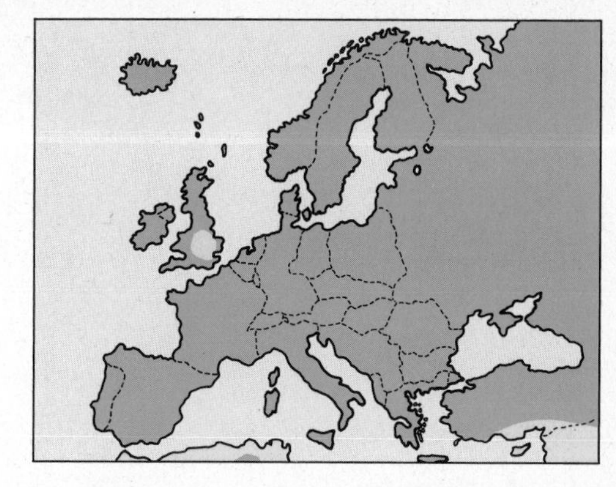

The Wheatear occurs in a variety of open habitats including moorland, dunes, dry grassy hillsides, alpine meadows, rocky scree slopes in high mountains, grassy areas on sea-cliffs and coastal islands. On migration it favours open grassy areas.

It feeds mainly on insects including beetles, grasshoppers, bees, Lepidoptera, etc, as well as spiders, centipedes and small earthworms. It is mainly terrestrial in habits, though often perches on the tops of bushes or on walls, fences and wires. It hops boldly on the ground but will make short flights to catch insects on the wing. The song is given from an exposed perch or in a bouncing song-flight.

The usual nest-site is in a hole in the ground, burrow or man-made object such as a pipe. It also nests in holes in walls or among rocks and heaps of stones. The nest is a large loose cup of dry grass, moss, plant stems and roots with a lining of fine grass, hair, wool or feathers. It is built mainly by the ♀, who also does most of the incubation, though the ♂ may assist with both nest-building and incubation. Both sexes feed the young.

The Wheatear is entirely migratory, wintering in open areas over a wide range in Africa south of the Sahara. It is a strong migrant and the Greenland race can make an oversea crossing of 2,500 km.

Migration: Departs breeding areas from late July to October, mainly August–September, returning in early March to April with passage continuing into early June in the north.

Length:	15 cm
Wing length:	8.9–9.9 cm (Greenland race larger)
Weight:	30 g
Voice:	Song: loud mixture of warbling and rattling noises. Calls: 'chack-chack' and 'weet-chack chack'
Breeding period:	Early April, May, June. 1, rarely 2 broods per year
Size of clutch:	5–6 (4–8) eggs
Colour of eggs:	Very pale blue, rarely with a few rufous spots at the blunt pole
Size of eggs:	21×15.5 mm
Incubation:	14 days, beg. from complete clutch
Fledging period:	Nidicolous; at 15 days the young leave the nest though not able to fly properly

♀, Greece, 26.5.1978 (Li)

Fledged young bird, Greece, 26.5.1978 (Li)

Öland, Sweden, 1978 (Schu)

) Ad. ♂, Småland, Sweden, 27.6.1969 (Chr)

Black-eared Wheatear

(Oenanthe hispanica)

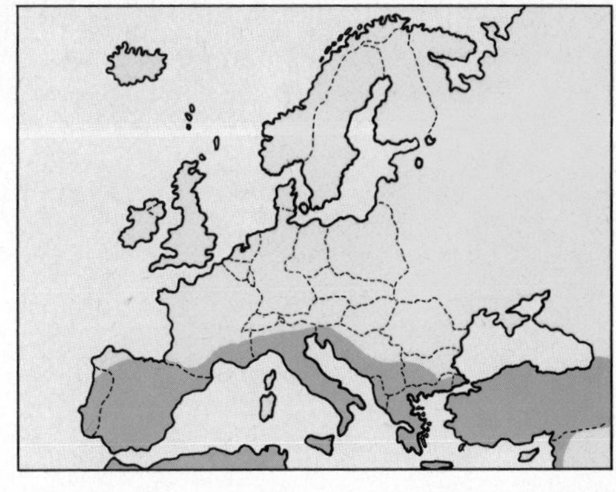

The Black-eared Wheatear inhabits warm, dry open country such as rocky hillsides, stony plains, Mediterranean heathlands, vineyards and fields with stone walls. It does not occur at such high altitudes as the Wheatear. Two plumage types exist, one with black ear-coverts and the other with a black throat as well. The diet is almost exclusively insects, mainly taken on the ground, and small molluscs and some seeds. The behaviour is much like that of the Wheatear.

The usual nest-site is in a hollow among stones or in low vegetation. The nest is a shallow cup of dry grass and moss, lined with hair or fine roots. It is built by the ♀ who also incubates alone. Both parents feed the young.

The Black-eared Wheatear is migratory, wintering in a narrow strip across Africa south of the Sahara.

Migration: Departs breeding areas in September–October, returning in March–April.

The **Pied Wheatear** (*Oenanthe pleschanka*) breeds in south-east Europe in countries bordering the Black Sea. It is an eastern ecological substitute of the Black-eared Wheatear. It occurs in habitat similar to that species but prefers more grassy open areas such as steppe, deep gullies and river banks. Its habits and breeding biology are much as in the Black-eared Wheatear, though it makes more use of observation-perches on bushes and small trees. It is entirely migratory, wintering in East Africa.

Migration: Departs breeding areas in September–November, returning in April–May.

Length:	14–15 cm
Wing length:	8.1–9.6 cm
Weight:	19–23 g
Voice:	Song: 'schwer-schwee-schwee-oo'; also 'plit' and rasping notes
Breeding period:	End of April, May, June, July. 1–2 broods per year; replacement clutch possible
Size of clutch:	4–5 (–6) eggs
Colour of eggs:	Blue-green with fine or coarser rufous spots, mainly round the blunt pole
Size of eggs:	24×15 mm
Incubation:	13–14 days, beg. from complete clutch
Fledging period:	Nidicolous; leaving the nest at 14 days, not yet able to fly

black-throated form, Spain, 1979 (Schu)

Yugoslavia, June 1969 (Li)

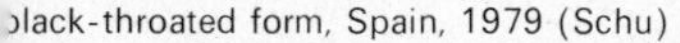

*) ♂, black-eared form, Macedonia, Yugoslavia, 1969 (Li)

Rufous Bushchat (Cercotrichas galactotes)

The Rufous Bushchat occurs in dry scrub, orchards, gardens, roadside hedges and clumps of bushes with open patches.
It feeds on a variety of insects, including grasshoppers and other large species, as well as spiders and earthworms. Much of the food is obtained on the ground though it will also take food from bushes and low shrubs. It is an active bird and often perched in exposed sites such as the top of bushes, cactus plants and rocks.
The nest-site is usually in a low bush, cactus plant or hedgerow. The nest is a rather untidy cup of plant stems, grass and other plant material, lined with hair, wool, feathers and fine roots. Both sexes build the nest but the ♀ incubates alone, fed by the ♂. Both parents feed the young. The species is migratory, wintering mainly in tropical East Africa.
Migration: Departs breeding areas in August–September returning in late April to May.

Black Wheatear (Oenanthe leucura) The Black Wheatear is found in arid rocky habitats. It occurs in mountainous areas up to 2,500 m and on sea-cliffs and low-lying desert areas. It closely resembles other Wheatears in diet and habits. The nest-site is in a hole or crevice in rocks or stony ground. The nest is a large, bulky cup of grasses, plant stems and other plant material, lined with hair, wool and feathers. Incubation is mainly or entirely by the ♀ and both parents feed the young.
The Black Wheatear is resident and sedentary, though some altitudinal movements may take place in winter and occasionally vagrants appear outside the breeding areas.

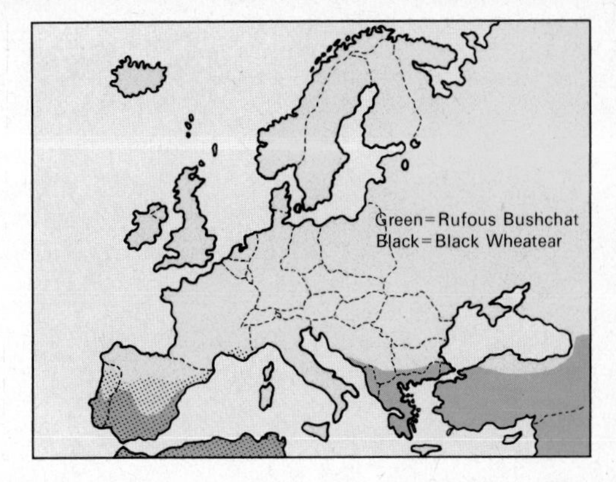

Rufous Bushchat	
Length:	15 cm
Wing length:	7.9–9 cm
Weight:	22 g
Voice:	'Tek-tek-tek'. Song: Short, curtailed verses, similar to Skylark
Breeding period:	May, June, July. 1–2 broods per year
Size of clutch:	4–5 (2–3) eggs
Colour of eggs:	Whitish, with fine or coarser spots in various shades of brown; zoned at blunt pole
Size of eggs:	22×16 mm
Incubation:	Unknown
Fledging period:	Nidicolous; period unknown

Black Wheatear	
Length:	18 cm
Wing length:	9.1–10.0 cm
Weight:	30 g
Voice:	Call: 'pee pee pee'. Song: mellow warble like Blue Rock Thrush, with some churning and rattling notes
Breeding period:	March, April, May. Probably 2 broods per year
Size of clutch:	4–5 (–7) eggs
Colour of eggs:	Bluish white, with sparse light and dark red spots, dense or zoned at the blunt pole
Size of eggs:	24'5×17.7 mm
Incubation:	16 days, beg. from complete clutch
Fledging period:	Nidicolous; leaving nest at 15 days, remaining with their parents till winter

k Wheatear ♂, Northern Spain, 1965 (Schw)

Black Wheatear ♀, Northern Spain, 1965 (Schw)

) Rufous Bushchat ♂, eastern race, Turkey, 22.5.1975 (Li)

Rock Thrush (Monticola saxatilis)

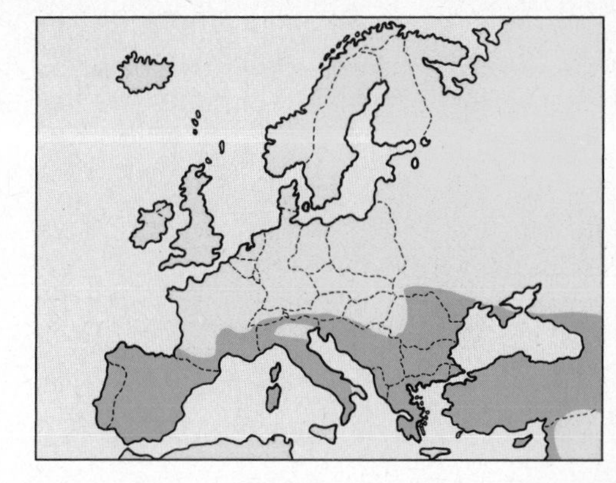

The Rock Thrush inhabits warm, dry rocky country, mostly in mountainous areas. It favours rocky slopes, cliffs, outcrops and ruined buildings. Though it occurs in some lowland areas it reaches a height of 2,500 m in Europe and much higher elsewhere in the range.

The diet is a variety of insects, mostly large species of beetle, Lepidoptera, etc. and spiders, small molluscs, worms and some berries. It spends most of the time on the ground, hopping amongst stones and boulders in search of food. The ♂ has a song-flight, rising with tail spread and gliding slowly back to a perch. It also sings from exposed rocks or on buildings.

The nest-site is a hole or crevice among stones or in a rock face. Sometimes holes in walls or ruined buildings are used and, exceptionally, in a tree. The nest is a cup of grasses, roots, plant stems and moss, lined with fine plant material. It is built by the ♀ who also incubates alone. Both parents feed the young.

The Rock Thrush is entirely migratory, wintering mainly in east and central Africa, where it is found among trees as well as rocky areas. The breeding range in Europe has contracted in the present century, probably due to climatic change.

Migration: Departs breeding areas in September–October, returning in late March to early May.

Length:	19 cm
Wing length:	11.4–12.9 cm
Weight:	65 g
Voice:	Call: 'chak-chak'. Song: a fluty warble, sometimes imitative
Breeding period:	May, mid-June. 1, rarely 2 broods per year. Replacement clutch possible
Size of clutch:	5 (4–6) eggs
Colour of eggs:	Blue-green, more intense than Blue Rock Thrush, rarely with very sparse fine red-brown spots
Size of eggs:	26×19.5 mm
Incubation:	14 days, beg. from complete clutch
Fledging period:	Nidicolous; leaving nest at 15 days

♂, Macedonia, Yugoslavia, 1971 (Li)

♂ in non-breeding plumage, Kenya, January 1977 (Wo)

South Tyrol, 1977 (Schu)

) Ad. ♀, Macedonia, Yugoslavia, 1971 (Li)

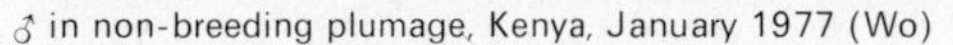

Blue Rock Thrush (Monticola solitarius)

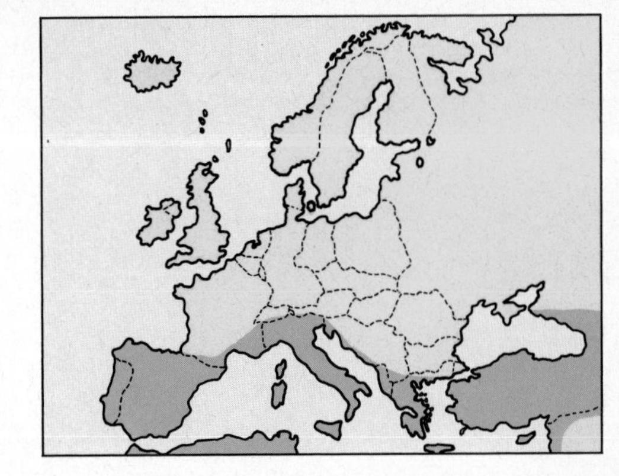

In Europe the Blue Rock Thrush generally occurs at a much lower altitude than the Rock Thrush. It is found in the same rocky habitat as that species but includes sea-cliffs, rocky islands, and steppe-type lowlands. It also inhabits buildings and ruins more frequently. The diet is quite varied and includes insects, spiders, earthworms, small reptiles, berries and small fruits, as well as crustaceans, molluscs and marine worms in coastal areas.
The ♂ has a song-flight, rising almost vertically and gliding back to a perch. It will also sing from the tops of boulders or buildings. It is a rather solitary bird, rarely associating with others of its own kind.
The nest-site is in a hole among rock or in a wall or building, or under an overhang on a bank or low cliff. The nest is a large, rather loose cup of dry grass and moss, lined with finer plant materials. It is built by the ♀, who incubates alone. Both sexes feed the young.
The Blue Rock Thrush is mainly resident and sedentary in southern Europe but some migration takes place, with a few birds wintering in parts of tropical Africa (mainly the north-east) and in parts of the Middle East and North Africa.
Migration: Passage occurs in September–October and late March to April.

Length:	20 cm
Wing length:	12.2–13 cm
Weight:	70 g
Voice:	Song: loud and fluty, resembling Blackbird's. Call: 'tsee' and 'chuck', like Rock Thrush
Breeding period:	End of April, May, June. 1–2 broods per year. Replacement clutch possible
Size of clutch:	4–5 (–6) eggs
Colour of eggs:	Light blue-green, sometimes with a few fine pale reddish spots
Size of eggs:	28×20 mm
Incubation:	12–13 days, beg. from complete clutch
Fledging period:	Nidicolous; remaining 18 days in nest

ear nest, Macedonia, Yugoslavia, 11.5.1972 (Li)

Ludwigsburg bird-reservation, 1965 (Schw)

t) Ad. ♂, Macedonia, Yugoslavia, 11.5.1972 (Li)

Mistle Thrush (Turdus viscivorus)

The Mistle Thrush occurs in broad-leaved and mixed woodlands, open pine forest, parkland with large trees, gardens and open areas with scattered tall trees. In some areas it is found mainly in hilly or mountainous country.
The diet is varied and includes snails, earthworms, insects and spiders. About half the diet comprises berries and fruit, and it is known to kill the young of other birds and feed these to the young.
It feeds on the ground and in trees and bushes, preferring taller trees. It is bold and aggressive in defence of the territory and will chase away much larger birds. Outside the breeding season it is often seen in small flocks or family parties.
The nest-site is usually in the fork of a tree or bush, 2–10 m from the ground, though lower bushes are used on occasion. The nest is a large cup of grass, plant stems, roots and moss with some earth used as a binding material. It is built by the ♀, who also incubates alone. Both parents feed the young and the ♂ may continue to look after the young when they have left the nest, allowing the ♀ to begin a second clutch.
In the south of the range, including southern England, the species is mainly resident and sedentary. In other areas it is migratory, wintering in south and south-west Europe and parts of the Middle East.
Migration: Autumn passage from mid-September to November, returning in mid-February to April.

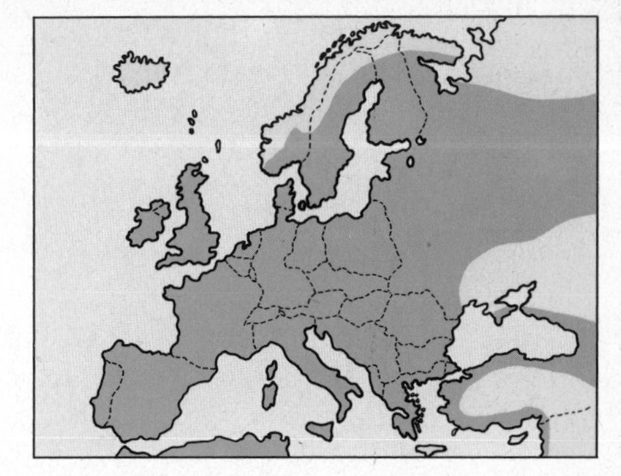

Length:	27 cm
Wing length:	14.5–16.0 cm
Weight:	120 g
Voice:	Call: 'tuc-tuc-tuc'. Song: Blackbird-like fluty notes, though less mellow; sings in all weathers from tall trees
Breeding period:	Late February to May, June, mid-July. 2 broods per year. Replacement clutch possible
Size of clutch:	4–5 (3–6) eggs
Colour of eggs:	Variable, bluish green to pale buff with spots of very varied shades of brown or purple
Size of eggs:	31×22.3 mm
Incubation:	14 days, beg. from last egg
Fledging period:	Nidicolous; leaving nest at 14–16 days, almost able to fly; flying at 20 days

, Bavaria, 10.3.1979 (Pf)

Fledged young bird, Bavaria, 1.6.1979 (Pf)

Bavaria, 8.5.1976 (Li)

) Ad., Bavaria, September 1976 (Pf)

Fieldfare (Turdus pilaris)

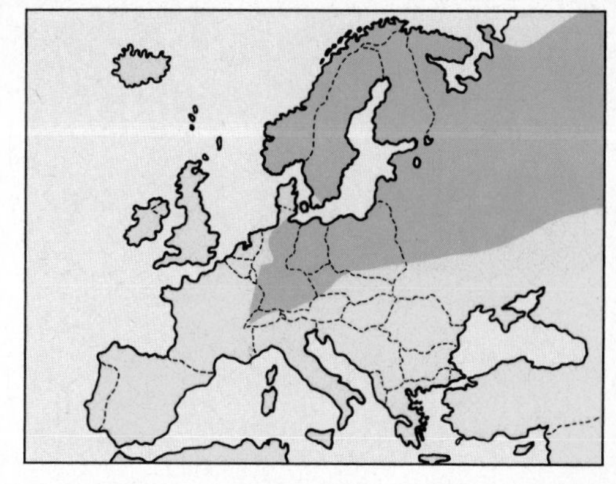

The Fieldfare is a typical bird of the northern taiga forests. In Europe it also inhabits mixed pine and birch woods, forest edges and clearing with grassy meadows, small clumps of trees in open areas, parks and gardens.
It feeds on a variety of insects, slugs, snails, spiders and berries. The latter make up much of the autumn and winter diet. It will also feed on fallen fruit such as apples and even resort to grain at times.
It perches freely in open places and feeds on the ground as well as in trees and bushes. Outside the breeding season it is highly gregarious and often met with in large flocks. It is aggressive in defence of the territory and will attack other birds and even human intruders.
The nest is usually placed in a rather open site, often a fork in a tree or in a post or stump. It will also nest on the ground or on steep banks. Often several nests are occupied in close proximity, forming a small colony.
The nest is a large cup of grasses, twigs, moss and other plant material, lined with a layer of mud and, on top of that, fine grass. The nest is built by the ♀ and she also incubates alone. Both parents tend the young, which may remain in family parties for some time after fledging, even on migration.
The Fieldfare is entirely migratory in the north of its range, wintering in western and southern Europe. The species has extended its range far to the south and west in recent times and now breeds regularly in Britain.
Migration: Departs breeding areas from late July to November, returning in March–May.

Length:	25–26 cm
Wing length:	13.5–15.3 cm
Weight:	100 g
Voice:	Call: 'cha-cha-cha-chack'. Song: weak and feeble, suggesting poor Blackbird with chattering notes
Breeding period:	April, May, mid-June. Often 2 broods per year. Replacement clutch possible
Size of clutch:	5–6 (3–8) eggs
Colour of eggs:	Light greenish blue, with dense red-brown spots, often forming a cap at the blunt pole
Size of eggs:	29×21 mm
Incubation:	13–14 days, beg. any time during laying
Fledging period:	Nidicolous; leaving nest at 14 days

at nest, Bavaria, 29.4.1972 (Li)

Nestlings, 2.6.1975 (Pf)

Bavaria, 6.5.1975 (Pf)

) Bavaria, 18.1.1977 (Li)

Ring Ouzel (Turdus torquatus)

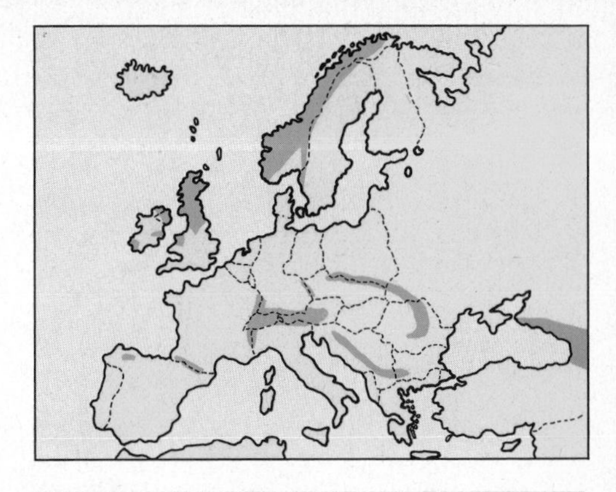

In the south of the range the Ring Ouzel is confined to high mountain forests and the upper scrub zone as well as in alpine meadows and by the side of mountain streams and rivers. In the north it occurs in moorland, upland valleys, rocky sea-coasts and islands, and in the birch scrub bordering tundra. The diet includes earthworms, insects, snails and a variety of berries and fruits.
In general the habits resemble those of the Blackbird but the Ring Ouzel rarely takes to cover, being a bird of open and wild terrain. It perches freely on exposed positions and is bold and aggressive in defence of the nest. On migration it is gregarious, though shyer and less approachable than other thrushes.
The nest-site is usually on the ground in hollow, rock-cleft or among low vegetation on a steep bank, ravine or rocky watercourse. In the south of the range the nest is usually placed in a tree or low bush, sometimes at a good height above ground. It is a bulky cup of grasses, plant stems, moss and leaves, mixed with some mud for binding and lined with fine grass. Both sexes share nest-building, incubation and care of the young.
The Ring Ouzel is almost entirely migratory, only some of the southern mountain population being resident. Most winter in southern Europe, North Africa and parts of the Middle East.
Migration: Departs breeding areas in September–October, returning in March–May.

Length:	24 cm
Wing length:	13.5–14.5 cm
Weight:	110 g
Voice:	Call: 'tac-tac-tac' and a trilling note. Song: repeated piping notes interspersed with chuckling sounds
Breeding period:	Mid-April–July. 1, sometimes 2 broods per year
Size of clutch:	4–5 (3–6) eggs
Colour of eggs:	Bluish green with sharply defined reddish-brown spots
Size of eggs:	30.5×21.5 mm
Incubation:	13–14 days, beg. after clutch complete
Fledging period:	Nidicolous; young leaves nest at *ca* 14 days

♂, Bavaria (Kü)

Nestlings, South Norway, 19.6.1965 (Sy)

Bavaria, 25.5.1976 (Li)

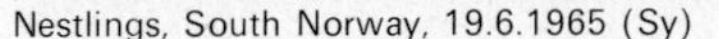

) ♂ at nest, Jämtland, Sweden, 19.6.1974 (Pf)

Blackbird (Turdus merula)

The Blackbird inhabits a vast range of differing habitats, from thick forest, both deciduous and coniferous, to parks and gardens in towns and cities, hedgerows in agricultural land, low cover near the tree-line in mountainous regions and desert oases. In western Europe it is a follower of human cultivation and habitation. The diet includes insects, earthworms, spiders, small snails, ants and a great variety of fruits and berries. It can be destructive to fruit crops in some areas.
It is rather skulking in habits and flies to cover readily. It will feed in open areas but rarely far from thick cover. Much of the food is obtained on the ground by turning over dead leaves. Only on migration is it gregarious.
The nest-site is usually in a tree or bush, placed in a fork in branches between 1 and 10 m from the ground. In western Europe it uses a variety of man-made habitats such as sheds, woodpiles, buildings, etc. Sometimes the nest is placed on the ground or in a tree-stump.
The nest is a large well-built cup of plant stems, grass, dead leaves, twigs and roots, lined with mud and then covered with fine plant material. It is built by the ♀, who also incubates alone. The ♂ may cover the eggs in the ♀'s absence but does not incubate. Both parents tend the young.
Over most of the range it is resident and sedentary, but northern populations migrate to winter in southern and western Europe including Britain and Ireland, north Africa and the Middle East.
Migration: Main movements are in October–November and March–April.

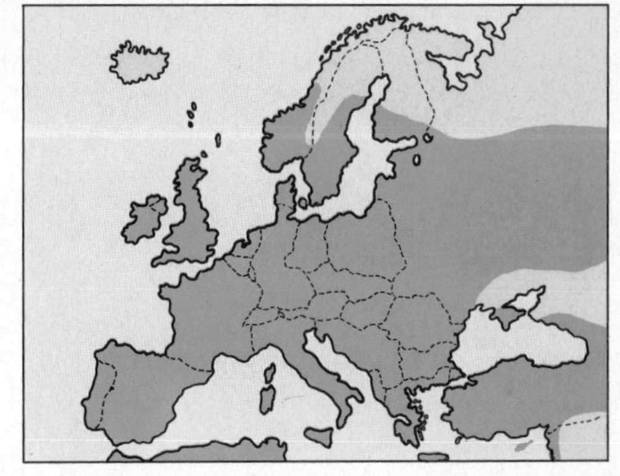

Length:	25 cm
Wing length:	11.8–13.5 cm
Weight:	100 g
Voice:	Calls: 'chink-chink', 'tsee' and 'chook-chook'. Song: rich and fluty
Breeding period:	End of March to end of July. 2–3 (–4) broods per year
Size of clutch:	4–5 (3–7) eggs
Colour of eggs:	Very variable. Bluish green to grey-green, evenly marked with washed-out red-brown spots
Size of eggs:	29×22 mm
Incubation:	12–15 days, beg. after clutch complete
Fledging period:	Nidicolous; leaving nest at 12–15 days, not yet able to fly well

ooding, Bavaria, 17.7.1972 (Pf)

Nestlings, Bavaria, 19.6.1979 (Pf)

Bavaria, 11.5.1971 (Pf)

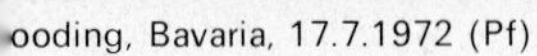

) Imm. ♂, Bavaria, November 1977 (Li)

Redwing (Turdus iliacus)

The Redwing is a characteristic bird of the northern taiga forests. It has extended its range into other habitats including birch- and alder-woods and even in parks and gardens in towns. It feeds on a variety of insects, slugs, snails and earthworms and berries of hawthorn, rowan, holly, etc.
Much of the food is obtained on the ground and it is less dependent on berries in winter than the Fieldfare, with which it often associates. It is essentially gregarious outside the breeding season and very large flocks occur on migration.
The nest-site is in a tree or bush, in a hollow tree-stump or among fallen branches. Ground-nests are also used, often on a steep bank or in broken ground. The nest is a substantial cup of twigs, grass, moss and lichen, sometimes with an inner lining of mud, in turn lined with fine grass. Nest-building and incubation are by the ♀ alone, with both parents feeding the young.
The Redwing is almost entirely migratory, wintering over much of southern and western Europe, parts of North Africa and Asia Minor. Two races are found in Europe, the nominate, which breeds on the continent, and the *coburni* race, which breeds in Iceland and occasionally in Britain. The Redwing has extended its breeding range to the south and west in recent years and has bred regularly in Britain since the mid-1960s.
Migration: Main autumn movements take place in September–October, with return passage in March to early May.

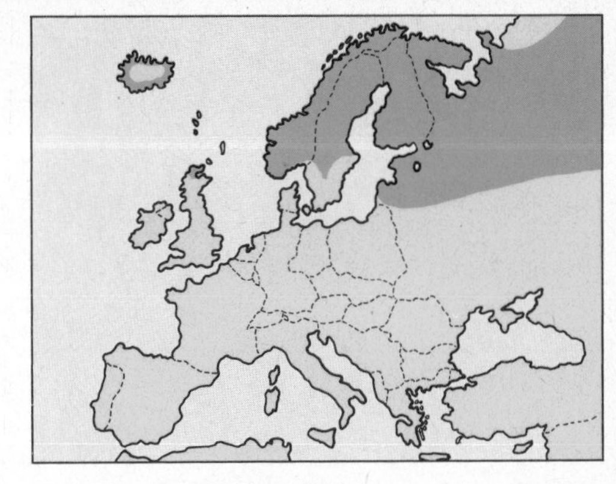

Length:	21 cm
Wing length:	11.3–12.5 cm
Weight:	60 g
Voice:	Flight-call: long-drawn-out 'seeeh'. Song: series of rising and falling fluty notes, 'trooi-trooi-trooi-trooi'
Breeding period:	Beginning of May to July. 1–2 broods per year
Size of clutch:	5–6 (2–8) eggs
Colour of eggs:	Bluish green with even fine red-brown patching
Size of eggs:	26×19.5 mm
Incubation:	12–15 days, beg. with penult. or last egg
Fledging period:	Nidicolous; leaving nest at 12–14 days

, Bavaria, 2.9.1975 (Li)

Ad. at nest, Norway, 22.6.1968 (Sy)

Finland (Hau)

t) Ad., Bavaria, 18.9.1975 (Li)

Song Thrush (Turdus philomelus)

The Song Thrush occurs in light woodland, mainly mixed or deciduous, and avoids the dense taiga forest. It is also found in parks, gardens, hedgerows and scrubby areas. In Scandinavia it does not occur in towns and cities, its place being taken by the Redwing.
It feeds largely on earthworms and snails, the latter being smashed against a stone 'anvil' which may be used repeatedly by one individual. Other food items include a variety of insects, spiders, centipedes, berries and fruit.
Though it feeds in the open it takes to cover more readily than other thrushes and rarely feeds far from suitable cover. It is not really gregarious except on migration, but small numbers associate where food is plentiful or among flocks of other thrushes.
The nest is usually well concealed against the trunk of a tree, or among creepers, in a hollow stump and sometimes on the ground or in a building. The nest is very characteristic and consists of a neat cup of grasses, twigs, roots, leaves and moss with a smooth inner lining of mud. It is built mainly by the ♀, who also does most of the incubation. Both parents tend the young.
In western and southern Europe the Song Thrush is largely resident and sedentary, though hard-weather movements take place. The northern and eastern populations are entirely migratory, wintering in southern and western Europe, North Africa and the Middle East.
Migration: Main autumn movements in September–October with return passage in March to early May.

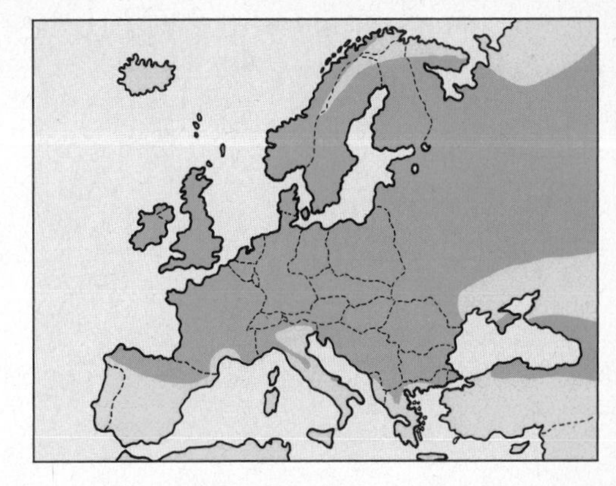

Length:	23 cm
Wing length:	11–12 cm
Weight:	70 g
Voice:	Flight call: 'tsip'. Song: loud and melodious with short verses, rhythmically repeated 2–3 times
Breeding period:	April–July. 2, sometimes 3 broods per year
Size of clutch:	4–5 (3–9) eggs
Colour of eggs:	Light blue with a few blackish-brown spots, generally concentrated at the blunt pole
Size of eggs:	27×20.5 mm
Incubation:	13–14 days, beg. after clutch complete
Fledging period:	Nidicolous; leaving nest at 12–15 days

collecting food, Bavaria, June 1972 (Li)

Nestlings, Bavaria, 19.5.1971 (Li)

Bavaria, 13.4.1969 (Li)

ɩ) Ad., Bavaria, 22.4.1971 (Pf)

Bearded Reedling or Tit

(Panurus biarmicus)

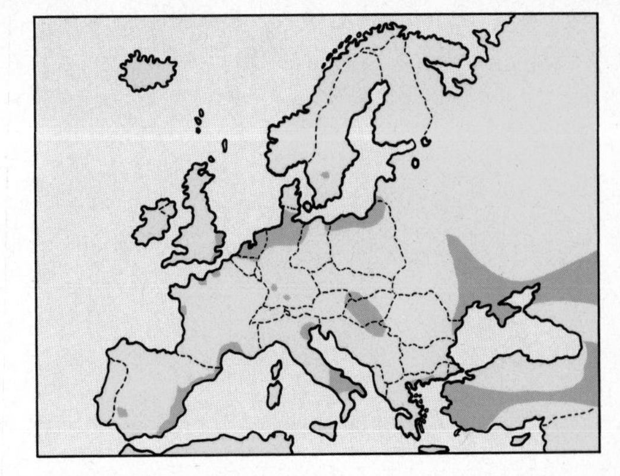

The Bearded Tit is found in extensive reed-beds, usually with phragmites the dominant plant. These reed-beds may be on fresh or brackish water, both inland and on the coast. The rather sporadic distribution is due to the restricted habitat and has been influenced by the draining of suitable marshes for agriculture.

In spring and summer the species is largely insectivorous, taking beetles, Lepidoptera, midges, etc. In winter it takes the seeds of marsh plants such as reed-mace, phragmites, etc. It keeps mostly to the cover of reeds where it moves with great agility. It also feeds on the ground, often in the open at the edge of the reed-bed. The flight is rather weak and whirring.

The nest is built among the stems of reeds or other plants, usually quite close to the ground or water level. The nest is a deep cup of stems and leaves of marsh plants, with some dry grass, and is lined with the flower-heads of reeds and a few feathers. Both sexes take part in nest-building, incubation and care of the young.

The Bearded Tit is largely resident and sedentary, and is often badly affected by hard winters, when the population may become very low or even extinct in some areas. When the population level is high some mass movements take place, sometimes with thousands of birds involved. These movements may range as far as 800 km and result in the colonisation of new areas, or the re-colonisation of suitable habitat where the previous population was wiped out by bad weather.

Length:	16.5 cm
Wing length:	5.7–6.2 cm
Weight:	14 g
Voice:	Calls: 'ping-ping' and 'ticc'; also twittering, wheezing and churring notes
Breeding period:	End of March to July. Up to 3, sometimes 4 broods per year
Size of clutch:	5–8 (–10) eggs
Colour of eggs:	Cream with very fine brown patches and scrawls
Size of eggs:	17.5 × 14 mm
Incubation:	11–13 days, beg. when clutch complete
Fledging period:	Nidicolous; leaving nest at 9–12 days

Austria, 1966 (Li)

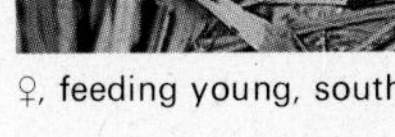

♀, feeding young, south Denmark, 27.6.1973 (Chr)

Greece, 16.6.1976 (Li)

ft) Ad. ♂, South Denmark, 27.6.1973 (Chr)

Long-tailed Tit (Aegithalos caudatus)

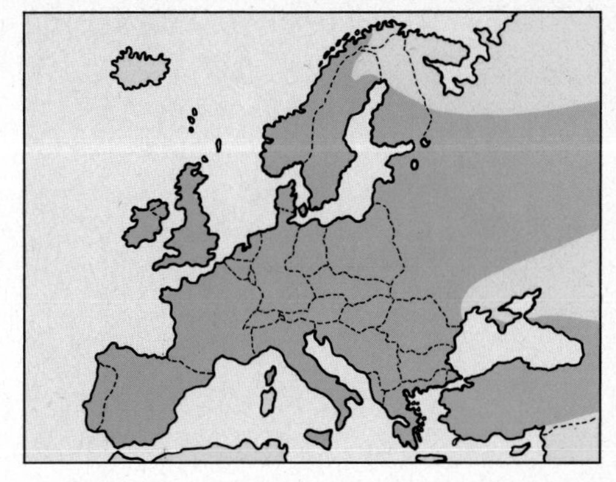

The Long-tailed Tit inhabits open woodland, mainly mixed or deciduous, woodland edges, thick hedgerows, copses and thickets. It does not occur in the dense spruce taiga. Three races breed in Europe including the white-headed form, which occurs in the north and east of the continent.
It feeds mainly on small insects and larvae, especially weevils, and Lepidoptera. It also takes some spiders, small seeds and buds. It feeds mainly in trees and bushes, rarely settling on the ground. It is acrobatic and very agile. It often associates with other Tits in feeding parties or occurs in unmixed family or larger groups. It rarely visits bird-tables or gardens.
The nest is usually in thick scrub or bushes or in tall trees, often conifers or evergreens. It is usually between 2 m and 6 m above ground but may be as high as 20 m. The nest is a large domed structure built of moss bound with spiders' webs or hair. The outside is decorated with lichen and the inside lined with feathers. It can take as long as three weeks to construct. Both sexes take part in nest-building and sometimes additional birds assist. Incubation is mainly by the ♀, who is fed by the ♂, while both parents tend the young.
The Long-tailed Tit is mainly resident and sedentary, though birds from the northern population sometimes disperse southwards in autumn and in some years may take part in irruptive movements on a larger scale.

Length:	14 cm
Wing length:	5.8–6.6 cm
Weight:	*ca* 9 g
Voice:	Repeated calls of 'tupp'; also 'si-si-si'
Breeding period:	End of March–June. 1 (–2) broods per year
Size of clutch:	8–12 (5–16) eggs
Colour of eggs:	Whitish with few or many delicate reddish spots, generally concentrated at blunt pole
Size of eggs:	14×11 mm
Incubation:	12–14 days, beg. before last egg laid
Fledging period:	Nidicolous; leaving nest at 15–16 days

ɪiped-headed race, Bavaria, 10.4.1979 (Pf)

Young birds, Hessen (Tö)

Lower Saxony, 27.4.1978 (Sy)

ft) White-haired race building nest, Bavaria, 1969 (Li)

Penduline Tit (Remiz pendulinus)

The Penduline Tit is found in marshy areas, beside lakes and rivers and in wooded steppe. It occurs in dense bushes and thickets with tamarisk, willow, brambles, etc.

It feeds on small insects taken from the extremities of twigs in trees and bushes or among reeds and reed-mace. The manner of feeding is more like that of a Warbler than a Tit. It also takes some small seeds, particularly in winter. It will often pick at the flower-heads of plants in the manner of a Goldfinch.

The nest is a remarkable structure, suspended from twigs near the tip of a branch overhanging the water. It is a purse-shaped construction with an entrance tunnel, made of plant fibres from reeds, grass and nettles interwoven with plant down from reeds, willow, poplars and cat's-tails. In some areas the nests are sited in reed-beds and built mainly from reeds and grass fibres. Both sexes take part in nest-building which can take about 2 weeks. Incubation is by the ♀ alone with both parents tending the young.

The Penduline Tit is mainly resident and sedentary, though some erratic dispersal takes place in autumn and winter. Sometimes this results in pairs breeding well outside the normal breeding range. In recent times there has been a tendency to extend the range westwards.

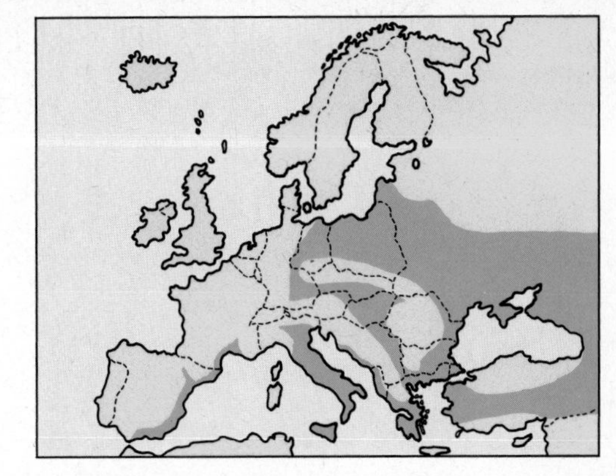

Length:	11 cm
Wing length:	5.7 cm
Weight:	10 g
Voice:	A low, long-drawn-out 'tseeh', resembling Reed Bunting; also 'tsi-tsi-tsi'
Breeding period:	Mid-April to August. 1–2 broods per year
Size of clutch:	6–7 (5–10) eggs
Colour of eggs:	Matt, pure white
Size of eggs:	15.5×10.5 mm
Incubation:	12–14 days, beg. after clutch complete
Fledging period:	Nidicolous; leaving nest at 15–18 (–20) days, remaining in family party for some weeks

ing ♂, Greece, 10.6.1976 (Li)

Ad. ♂, European Turkey, 15.5.1966 (Fe)

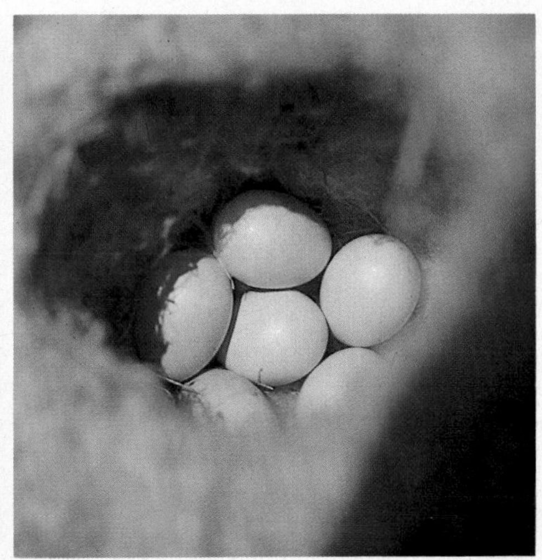

Nest opened by rats, Greece, 10.6.1979 (Li)

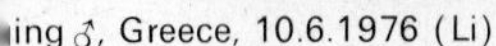

♀ at nest, Greece, 11.6.1979 (Li)

Crested Tit (Parus cristatus)

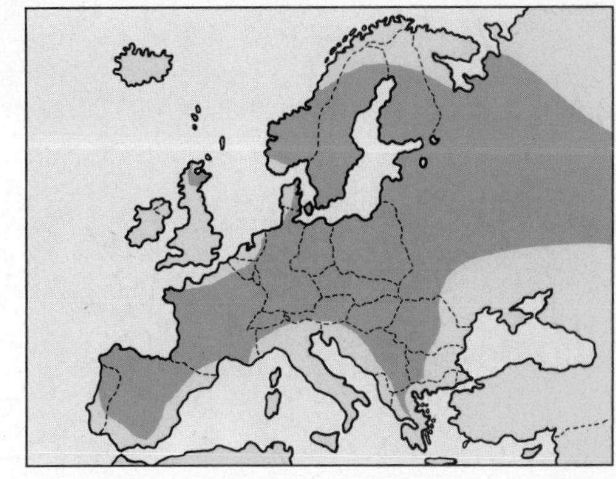

The Crested Tit is found almost exclusively in coniferous forests of mainly pine or spruce. It also occurs in some mixed woodland and montane beechwoods, though usually near the coniferous zone. It occurs in lowland areas, but mainly in upland or mountainous regions. Due to its habitat preference it is the least adaptable of the European tits.
The main food is insects and larvae, mostly of small Lepidoptera, as well as seeds from ripe pine-cones. It searches for food in the higher branches of trees and rarely joins the mixed flocks of other tits. In some areas it comes to bird-tables.
The nest-site is in a hole excavated by the birds in a rotten tree-stump or dead tree. It sometimes uses an old nest of larger birds or a squirrel drey. It takes readily to nest-boxes where natural sites are scarce. The nest is a cup of moss and lichens inside the hole and lined with hair wool or spiders' webs. The nest is built by the ♀ who also excavates the nest-hole. She incubates alone, fed by the ♂. Both parents feed the young. At first the ♀ feeds them with food brought by the ♂. Later both birds bring food, but again the ♀ usually feeds the young. The ♂ may take over feeding if the ♀ begins a second brood.
The Crested Tit is resident and sedentary throughout its range. Only exceptionally does it wander away from the normal habitat.

Length:	11.5 cm
Wing length:	5.8–6.2 cm
Weight:	11 g
Voice:	Humming 'choo-urr' or 'tze-zee-zee'
Breeding period:	Beginning of April to end of June. 1–2 broods per year
Size of clutch:	6–7 (5–9) eggs
Colour of eggs:	Matt white with rusty red spots in a zone at the blunt pole
Size of eggs:	16×12.5 mm
Incubation:	13–15 days, beg. when clutch complete
Fledging period:	Nidicolous; leaving nest at 17–21 days

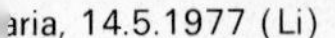

aria, 14.5.1977 (Li)

Bavaria, 10.5.1977 (Li)

) Ad. at nesting hole, Bavaria, 14.5.1977 (Li)

Marsh Tit (Parus palustris)

The Marsh Tit forms a sibling species with the Willow Tit. It occurs in deciduous woodland, mainly beech and oak usually with a dense undergrowth. It is also found in riverine woods of willow and poplar, copses, thickets, parks and large gardens. It feeds on insects, mainly beetles, and a variety of seeds, berries and nuts. Like the Willow Tit it tends to feed more in lower growth than in the canopy. Despite its name, it has no special attachment for marshy places though it sometimes occurs in damp situations. The nest is in a natural hole in a tree or stump or one excavated or modified by the ♀. It will also nest in holes in walls and takes readily to nest-boxes. The nest is a cup of moss and fine plant materials lined with hair and feathers. It is built by the ♀ who also incubates alone, being fed by the ♂ when brooding. Both parents tend the young who remain with the parents for a week or so after fledging.
The species is largely resident and sedentary in Europe.

The **Sombre Tit** (*Parus lugubris*) occurs in south-east Europe. It is similar in appearance to the Marsh or Willow Tit but is considerably larger. It is found in montane woods of oak, beech or pine as well as in vineyards, orchards and open broad-leaved woods beside water. Its diet, habits and breeding biology resemble those of the Marsh Tit. It is resident and sedentary throughout its range. In some areas it overlaps with the Marsh and Willow Tits but it is not known whether there is inter-specific competition.

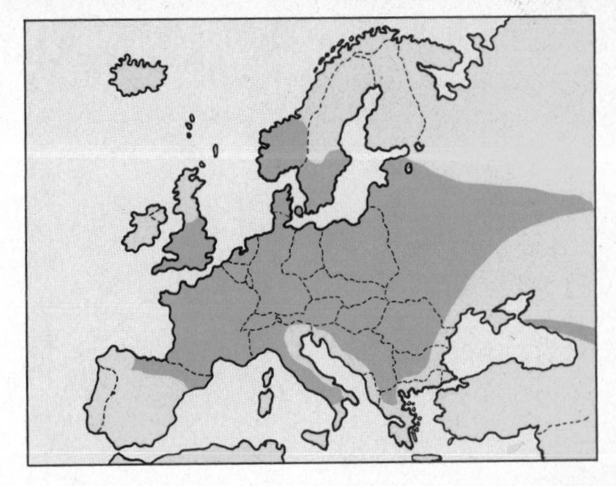

Length:	11.5 cm
Wing length:	5.9–6.5 cm
Weight:	11 g
Voice:	Call: 'pitchuu' or a clamorous 'tcha-tcha-tcha-tcha'. Song: a bubbling 'tsyetsyetsye'
Breeding period:	End of April to beginning of June. 1, sometimes 2 broods per year
Size of clutch:	7–9 (5–12) eggs
Colour of eggs:	Smooth, white ground with red-brown spots, ranged in a zone round the blunt pole
Size of eggs:	16.5 × 12.5 mm
Incubation:	13–15 days, beg. before last egg laid
Fledging period:	Nidicolous; leaving nest at *ca* 17–20 days

'aria, 1972 (Br)

Nestlings in nesting-box, Småland, Sweden (Chr)

Bavaria, 1975 (Rei)

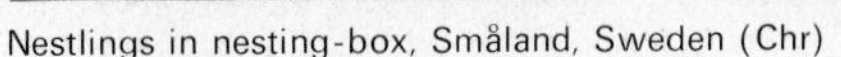

:) Ad., Bavaria, 12.6.1979 (Pf)

Willow Tit (Parus montanus)

The Willow Tit prefers damp woodlands such as alder, willow and birch as well as coniferous forests and mixed woodland. The habitat overlaps with that of the Marsh Tit in some areas but the Willow Tit generally prefers swampy places, or montane woodland, in central Europe. The diet consists mainly of small insects and differs from that of the Marsh Tit only by the fact that it contains fewer seeds.
The nest-site is a hole in a dead tree or stump which is excavated by the ♀. Sometimes a natural hole or one made by a woodpecker is used. The hole is lined with wood-chips and a little moss, sometimes lined with hair or a few feathers. Incubation is by the ♀ alone, being fed by the ♂, while both parents tend the young.
The Willow Tit is largely resident and sedentary except in the extreme north of the range, where it disperses in winter, sometimes in quite large flocks.

The **Siberian Tit** (Parus cinclus) inhabits the northern taiga forests and mixed or broad-leaved woodland along river valleys on the edge of the taiga zone. The food is mainly small insects and spiders taken in the tops of trees. In winter a large number of pine- and cedar-seeds are taken. The nest-site is a natural hole, or abandoned woodpeckers' hole in a tree, also in decaying tree-stumps. The ♀ may partly excavate holes in rotten wood. The nest is built by the ♀ from moss and lined with hair or feathers. Incubation is by the ♀ alone, being fed by the ♂, while both parents feed the young.
The species is resident and sedentary, though sometimes it wanders in winter to the south of the breeding range.

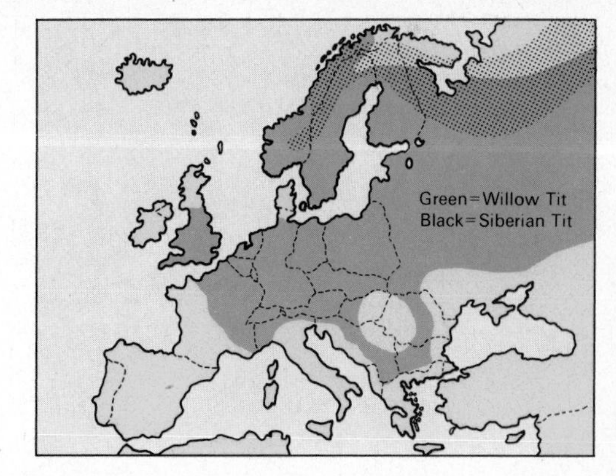

Length:	12–13 cm Willow Tit
Wing length:	5.6–6.2 cm
Weight:	10–13 g
Voice:	Long-drawn-out 'zi-zurr-zurr-zurr' also 'eez-eez-ezz'
Breeding period:	Eng of April to June. 1 brood per year, rarely 2
Size of clutch:	7–8 (5–13) eggs
Colour of eggs:	Glossy white with many or few red-brown patches, especially at blunt pole
Size of eggs:	15.5×12.2 mm
Incubation:	13–16 days, beg. when clutch complete
Fledging period:	Nidicolous; leaving nest at 17–19 days

w Tit, Hessen, West Germany, 1963 (Pf)

Siberian Tit, Sweden (Sw)

Length:	13–14 cm Siberian Tit
Wing length:	*ca* 7 cm
Weight:	*ca* 12–14 g
Voice:	Similar to Willow Tit; also long-drawn-out, repeated calls of 'eeez'
Breeding period:	1 brood per year. Replacement clutch possible
Size of clutch:	7–9 (6–10) eggs
Colour of eggs:	White with regular red-brown or pale red spots, increasing towards the blunt pole
Size of eggs:	16.5×12.5 mm
Incubation:	14–15 days, beg. when clutch complete
Fledging period:	Nidicolous; leaving nest at 19 days

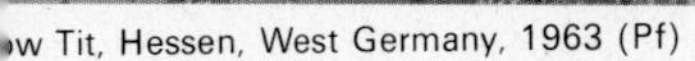

Willow Tit, at breeding-hole, Bavaria 1967 (Li)

Blue Tit (Parus caeruleus)

The Blue Tit breeds in deciduous woodland, copses, thickets, hedges, gardens and parks, and in trees and bushes in cultivated areas. It occurs in both lowlands and mountains, and is frequent in towns and cities.
The diet includes a variety of small insects, the young being fed mainly on the larvae of Lepidoptera, spiders, millipedes, seeds, berries and buds (it sometimes causes damage to fruit trees) and in some urban areas has developed a habit of taking cream from milk-bottles after puncturing the top. The species is a frequent visitor to bird-tables and artificial feeding-stations. The nest-site is a hole in a tree, wall or in nest boxes. The nest is a cup of moss, wool, leaves and spiders' webs, lined with hair. It is built by the ♀ who also incubates alone, being fed by the ♂. Both parents feed the young. The Blue Tit is mainly resident and sedentary, though most northerly birds move south in winter.

Azure Tit (*Parus cyanus*) This species is found in eastern Europe and the range overlaps with that of the Blue Tit. The species sometimes hybridises with the Blue Tit on the edge of its range. The habitat is mainly willow or poplar bushes along rivers, bushes and low woodland near open fresh water as well as open steppe woodland and deciduous woods in montane areas. The breeding behaviour, diet and habits are much as for the Blue Tit. The species is resident and sedentary and rarely occurs outside the breeding range, which appears to be decreasing eastwards.

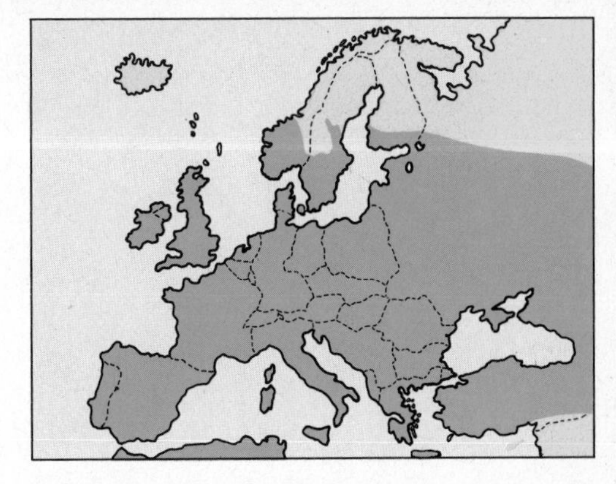

Length:	11–12 cm
Wing length:	6.0–6.8 cm
Weight:	11 g
Voice:	'Tsee-tsee-tsee' or 'tsee-tsee-tseet-sit'. Warning call a screaming 'tserrre-tette'
Breeding period:	End of April to July. 1–2 broods per year
Size of clutch:	7–14 (5–16) eggs
Colour of eggs:	White ground with sparse reddish spots
Size of eggs:	16×12 mm
Incubation:	13–15 days, often beg. during laying
Fledging period:	Nidicolous; leaving nest at *ca* 16 days

., Bavaria, 6.1.1978 (Pf)

Young bird, June 1979 (Pf)

Bavaria, 6.5.1977 (Li)

ft) Ad., Bavaria, 29.5.1979 (Pf)

Great Tit (Parus major)

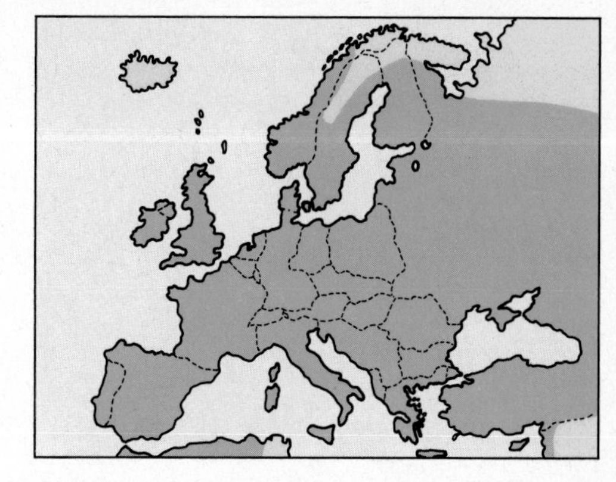

The Great Tit inhabits mixed and deciduous woodland of all types, copses, orchards, scattered trees in parkland, gardens, hedgerows, and trees in agricultural land. It occurs from sea-level to near the limit of the tree-zone in mountains.
The diet includes a variety of insects and their larvae, spiders, small earthworms, molluscs, seeds, berries, peas and buds. It feeds in the tops of trees, in lower growth and on the ground. It comes readily to bird-tables and other artificial feeding-stations, particularly in winter. In autumn and winter it often gathers in small flocks which roam through woodland, etc.
The nest-site is in a hole in a tree, or in a wall or among stones. It also uses the old nests of large birds and squirrel dreys, and will take to nest-boxes. The nest is a cup of moss, grass, lichens and roots lined with hair and plant down. It is built by the ♀ who also incubates alone, being fed by the ♂. Both parents tend the young.
The Great Tit is mainly resident and sedentary, but birds from northern populations tend to move southwards in autumn. In periods of hard weather some continental birds reach Britain.

Length:	14 cm
Wing length:	7.0–7.8 cm
Weight:	20 g
Voice:	Variety of sounds, e.g. 'tui-tui-tui', 'pink, pink'. Song: 'tee-chu-tee-chu-tee-chu'
Breeding period:	End of April to July. 1–2 broods per year
Size of clutch:	8–10 (5–16) eggs
Colour of eggs:	White, with reddish spots and scrawls
Size of eggs:	18×14 mm
Incubation:	13–14 days, beg. when clutch complete
Fledging period:	Nidicolous; leaving nest at 15–20 days

♂ at breeding-hole, Bavaria, 1978 (Li)

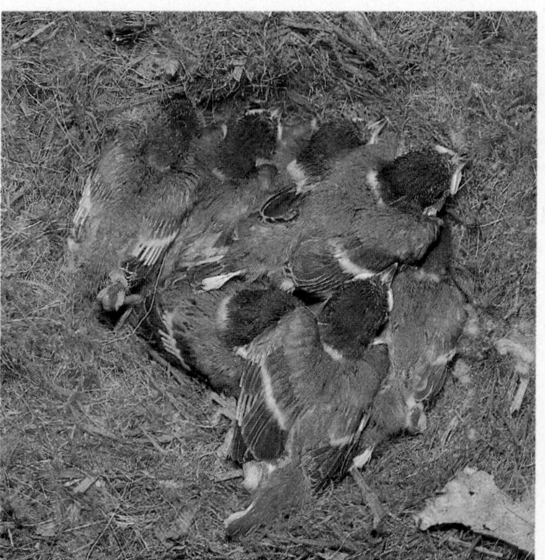

Nestlings, Bavaria, June 1977 (Pf)

Bavaria, May 1978 (Pf)

t) Ad. ♀, February 1978 (Pf)

Coal Tit (Parus ater)

The Coal Tit is typically found in coniferous woodland, but it also occurs in mixed and some deciduous woods, parkland with copses, gardens and cultivated areas with scattered trees. It occurs less often in pure pine-woods than the Crested Tit, preferring spruce and firs. It is found in both lowland and mountainous country. The diet is mainly small insects at all stages, including larvae and eggs as well as spiders, seeds, berries and nuts. It will come to bird-tables and artificial feeding-stations for fat or seeds. It feeds more on the trunks of trees than other tits, searching for food in the bark like a treecreeper. It is gregarious outside the breeding season and often occurs in small flocks or mixed with other species of tit.
The nest is usually in a hole in a tree, or wall, and sometimes in a hole in a bank or on the ground. The nest is a cup of moss and spiders' webs, lined with hair, plant down and feathers. Nest-boxes are frequently used, especially where natural sites are scarce. The nest is built by the ♀ who also incubates alone, fed by the ♂. Both sexes tend the young.
The Coal Tit is mainly resident and sedentary, though northernmost populations move south in winter. Some dispersal takes place in hard weather with some continental birds reaching Britain.

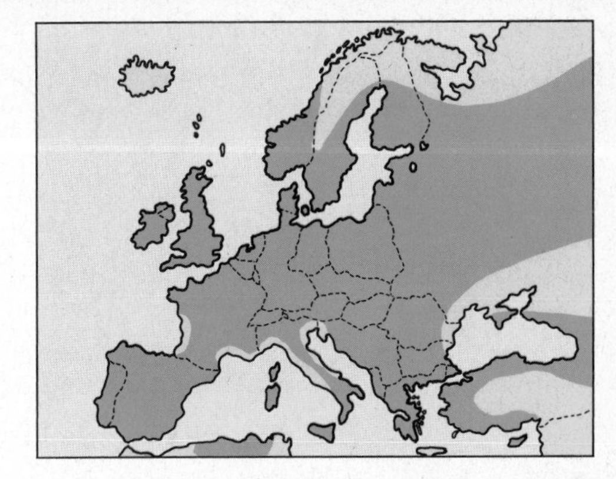

Length:	11 cm
Wing length:	5.9–6.3 cm
Weight:	8–9 g
Voice:	Call like Goldcrest, 'sissi-sissi-sissi'; also 'tsuu' and piping song: 'tee-chu-tee-chu . . .'
Breeding period:	Mid-April to July. 2 broods per year
Size of clutch:	7–10 (5–14) eggs
Colour of eggs:	White, with sparse reddish spots, increasing towards the blunt pole
Size of eggs:	15×12 mm
Incubation:	14–18 days, beg. before clutch complete
Fledging period:	Nidicolous; leaving nest at 16–19 days

‹dged young birds, Bavaria, 15.5.1977 (Li)

Nestlings in nest-box, Bavaria, 22.6.1978 (Pf)

Bavaria, 10.6.1978 (Pf)

) Ad., Bavaria, 21.6.1978 (Pf)

Rock Nuthatch (Sitta neumayer)

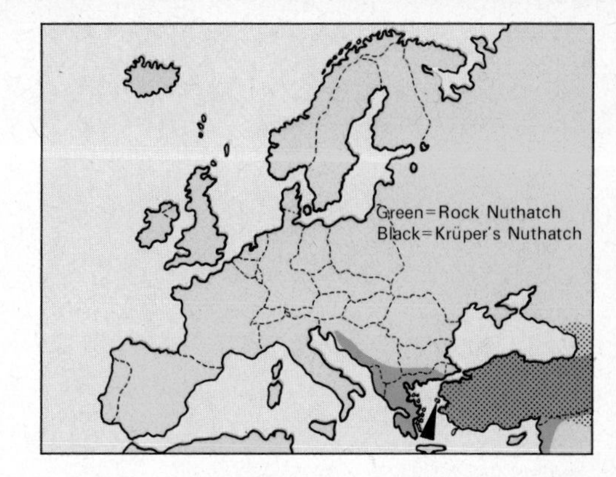

The Rock Nuthatch is found in sunny, warm rocky areas, often limestone cliffs or outcrops, from sea-level to over 1,500 m, often in completely bare areas, or at most with a low growth of scrub. In winter it will often occur in areas of low bushes and trees.
The food is mainly insects and spiders which are obtained on the ground, among stones and rocks, in crevices in rock-faces and also in bushes and trees. The species is not gregarious, but is commonly seen in pairs.
The nest is a rounded structure of mud with an entrance tunnel, built on a rock-face under an overhang or in a hollow or crack in the rock. The nest is built mainly by the ♀ and takes up to ten days to build. An inner cup of moss, hair, wool or feathers is provided. The incubation is by the ♀ alone and the young are fed by both parents. The species is resident and sedentary.

Krüper's Nuthatch (Sitta kruperi) This species is found only in the extreme south-east of Europe and in Asia Minor. It occurs in coniferous forests on hillsides or mountainous country. The diet is mainly insects which are obtained when searching the bark of trees and on outer twigs and branches. The nest is a hole in a tree, sometimes enlarged by the bird. The old nests of woodpeckers are used and sometimes a cavity behind the bark of a dead tree. The nest is a cup of moss, hair, wool and feathers, built by the ♀. The entrance to the hole is not plastered with mud. Little is known of the breeding biology or behaviour. The species is resident and sedentary.

Length:	14 cm Rock Nuthatch
Wing length:	8 cm
Weight:	24–33 g
Voice:	Falling 'tlwitlwitlwi' and 'sia, sia, sia'
Breeding period:	End of March to June. 1–2 broods per year
Size of clutch:	8–9 (6–10) eggs
Colour of eggs:	Cream with yellow-brown to reddish spots, concentrated at blunt pole
Size of eggs:	20.5×15 mm
Incubation:	14–15 days, beg. when clutch complete
Fledging period:	Nidicolous; fledging period unknown

Length:	12.5–13 cm Krüper's Nuthatch
Wing length:	*ca* 7 cm
Weight:	18–20 g
Voice:	Like Greenfinch, 'dwit'; also series of calls: 'kvee, kvee, kvee' and 'shwish'
Breeding period:	1 brood per year. Replacement clutch possible
Size of clutch:	5–6 eggs
Colour of eggs:	White with fine reddish-brown spots and patches, concentrated at blunt pole
Size of eggs:	17×13 mm
Incubation:	Unknown
Fledging period:	Unknown

k Nuthatch, Macedonia, Yugoslavia, 1972 (Li)

Krüper's Nuthatch ad., Turkey, 19.5.1975 (Li)

) Rock Nuthatch feeding young, Turkey, June 1974 (Pf)

Nuthatch (Sitta europaea)

The Nuthatch is found in deciduous and mixed woodland, parks, gardens, roadside trees, olive groves, etc. In the north it inhabits the taiga forests. Large old trees are the main requirement, providing suitable nest-sites and a wealth of food.

The diet includes a variety of insects and larvae of tree-dwelling species as well as spiders and small snails. In autumn and winter seeds and nuts make up the main part of the diet. The strong bill is able to cope with hazelnuts, acorns, beech-mast, etc. Nuts and large seeds are usually wedged in a crack in the bark of a tree, leaving the bird free to use the whole weight of the body behind the blows, the legs acting as hinges. It will also come to bird-tables and artificial feeding-stations for seeds and fat in winter.

The nest-site is in a hole in a tree, sometimes an old nest of a woodpecker or in a nest-box. The size of the entrance-hole may be reduced with a plaster of mud. The nest is lined with bark chips and dead leaves. The ♀ makes the lining but the ♂ may assist with plastering up the entrance. Incubation is by the ♀ alone, being fed by the ♂. Both parents tend the young.

The species is mainly resident and sedentary, though northern birds wander in winter.

The **Corsican Nuthatch** (*Sitta whiteheadi*) breeds only in Corsica, where it occurs in mountain pine-woods. It is similar to the Nuthatch but had a black crown and is much smaller. The diet, habits and breeding biology are very similar to those of the Nuthatch, though it does not plaster the entrance to the nest-hole. It is entirely resident.

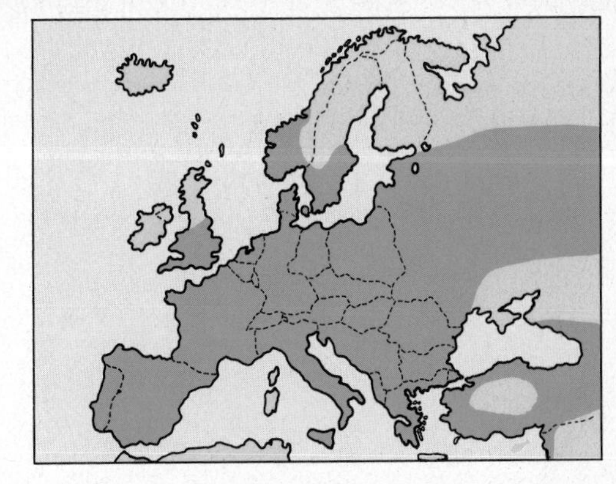

Length:	14 cm
Wing length:	8.0–8.8 cm
Weight:	*ca* 23 g
Voice:	Metallic 'chwit-chwit' also 'tsirrup' and 'pee' or 'chee'
Breeding period:	Mid-April to early June. 1, sometimes 2, broods per year
Size of clutch:	6–8 (5–12) eggs
Colour of eggs:	Milky white with fine reddish or brownish patches, concentrated at blunt pole
Size of eggs:	20×15 mm
Incubation:	14–17 days, beg. from complete clutch
Fledging period:	Nidicolous; young develop very slowly, and leave nest at 22–25 days

, Scandinavian race, Sweden, 13.5.1974 (Chr)

Brood in nesting-box, Amperaue, Bavaria, 20.5.1972 (Li)

Amperaue, Bavaria, 24.4.1973 (Li)

t) Central European race, Bavaria, 1975 (Li)

Wallcreeper (Tichodroma muraria)

The Wallcreeper is found on steep rock-faces in mountainous areas, often near the snow-line or near glaciers. In winter it is also found on the walls of castles and other large buildings. It occurs up to 4,500 m in Europe, and shares its habitat with the Alpine Chough. The food is mainly insects, particularly larvae, as well as spiders, centipedes and small snails. It feeds mostly on dark, damp rock-faces rather than sunny ones. The food is obtained from among rock-crevices, under stones, etc. The species climbs and flits about on the vertical rock but does not use the tail as a support, like Treecreepers. It is usually seen singly or in pairs, but small parties are sometimes seen.
The nest-site is in a cave, hole or rock-crevice, usually well hidden. The nest is a cup of grass, moss, roots and wool, lined with hair and feathers. It is built by the ♀ who also incubates alone. The young are fed by both parents.
The species is mainly resident and sedentary though it often descends to a lower altitude in winter and may occur in towns (e.g. Budapest, Vienna). Sometimes longer movements occur and the species has been recorded well away from the breeding area including Britain, Finland and Malta.

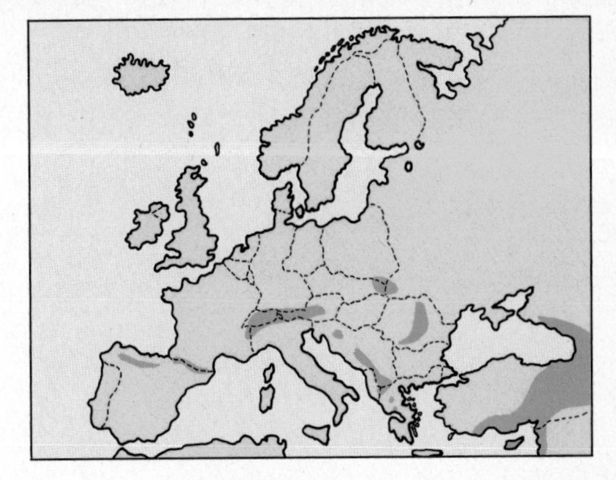

Length:	17 cm
Wing length:	9.3–11.0 cm
Weight:	20 g
Voice:	Piping call, 'tyoo'. Song: a melodious 'zee-zee-tiri-zwee'
Breeding period:	Mid-May to mid-June. 1 brood per year
Size of clutch:	4–5 (3) eggs
Colour of eggs:	White with a few red-brown spots, esp. at blunt pole
Size of eggs:	20.9×15 mm
Incubation:	18–19 days, beg. from complete clutch
Fledging period:	Nidicolous; leaving nest at 21–26 days

eding young, Tyrol, Austria (Aich)

Nestlings, Tyrol (Aich)

Tyrol (Aich)

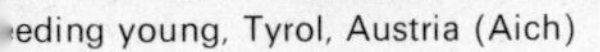

) ♂ displaying (Aich)

Treecreeper (Certhia familiaris)

The Treecreeper forms a sibling species with the Short-toed Treecreeper in Europe. The Treecreeper is found in coniferous, mixed and some deciduous woodland (the latter in places where the short-toed Treecreeper does not occur). It prefers more upland or montane forests to the Short-toed species. The diet is mainly small insects and larvae, spiders and woodlice, as well as a few seeds. The food is obtained by searching cracks in the bark of trees and the species runs rapidly up and around tree-trunks. It uses its tail as a support like a woodpecker and rarely descends head-down like a Nuthatch. It is not gregarious and is seldom seen in a group other than family parties, but often associates with mixed feeding flocks of Tits in winter.

The nest is concealed behind loose bark on a tree or in a crevice, and is a loose cup of twigs, moss, roots and plant material, lined with feathers, wool or fine bark débris. Both sexes take part in nest-building. The ♀ incubates alone and the young are tended by both parents.

The Treecreeper is mainly resident and sedentary, though some movement of northern birds takes place in winter, with some of the Scandinavian form occasionally reaching Britain.

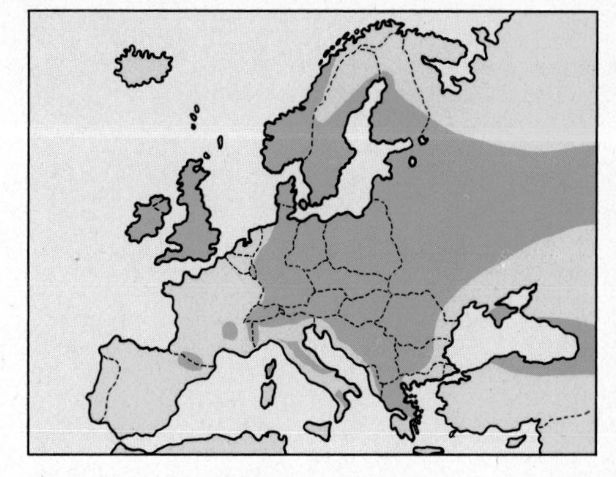

Length:	13 cm
Wing length:	5.9–6.7 cm
Weight:	9 g
Voice:	Thin, high 'tsu' or 'tsit'. Song: thin and high pitched, like Goldcrest
Breeding period:	End of April to June. 2 broods per year
Size of clutch:	5–6 (3–9) eggs
Colour of eggs:	White ground with fine rusty spots, esp. round blunt pole
Size of eggs:	15.5×12 mm
Incubation:	14–15 days, beg. from complete clutch
Fledging period:	Nidicolous; leaving nest at 14–17 days

…l., North Rhine–Westphalia, June 1973 (Sieb)

Fledged young birds, Bavaria (Kü)

Bavaria, 1969 (Li)

…ft) Ad., Baden-Württemberg, 1974 (Schw)

Short-toed Treecreeper

(Certhia brachydactyla)

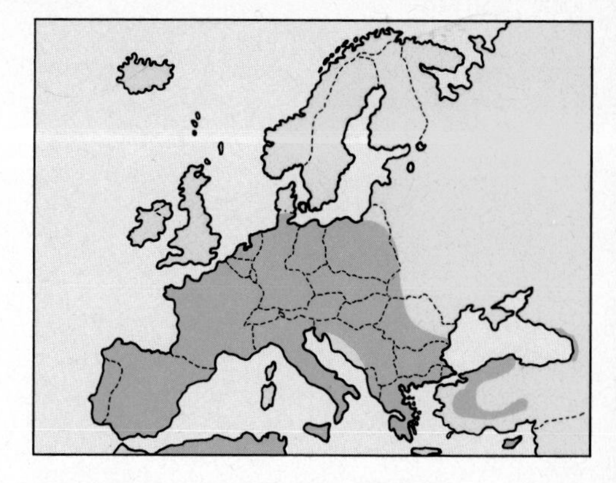

The Short-toed Treecreeper is found mainly in deciduous woodland, particularly in open areas and at a lower altitude than the Treecreeper. In areas where the Treecreeper does not occur it will inhabit coniferous and even montane woodlands. In areas where both species occur the difference in habitat and altitude are fairly marked but both can occupy the other's favoured habitat where one species does not occur.

The diet, habits and breeding biology are very similar to those of the Treecreeper, though the Short-toed Treecreeper will nest more readily in holes in stone walls or cracks or holes in fence-posts.

The species is mainly resident and sedentary, though in recent years it has occurred as a vagrant in southern England and has been suspected of breeding. It is the only form occurring in the Channel Islands.

The two species are extremely similar in appearance, and the song and call notes provide the best method of distinguishing them.

Length:	*ca* 13 cm
Wing length:	6.5 cm
Weight:	9 g
Voice:	Song: loud and rhythmic, 'teet-teet teeteroititt'. Calls: 'zeet' or 'srrieh', like Dunnock
Breeding period:	Mid-April to July. 2 broods per year
Size of clutch:	6–7 (4–7) eggs
Colour of eggs:	White with rusty spots, concentrated at blunt pole. Spots more pronounced than with Treecreeper
Size of eggs:	16 × 12 mm
Incubation:	15–16 days, beg. when clutch complete
Fledging period:	Nidicolous; leaving nest at 16–17 days

, Bavaria, 18.4.1978 (Pf)

Nestlings, Bavaria, 1.5.1976 (Pf)

Bavaria, 20.4.1976 (Pf)

ɫ) ♂ and ♀ at nest, Bavaria, 28.4.1978 (Pf)

Corn Bunting (Miliaria calandra)

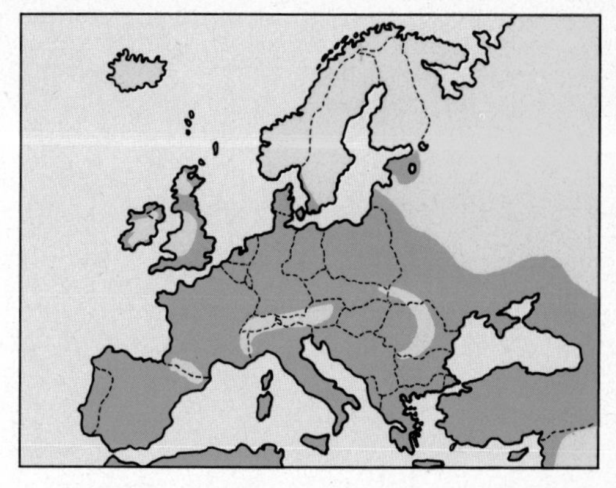

The Corn Bunting is a bird of open country such as grasslands, heath, agricultural land and wasteland. It prefers dry areas with sparse cover, though it will occur in Mediterranean *maquis* scrub. It feeds mainly on seeds of low plants, shoots and buds as well as a variety of insects, spiders and centipedes. Most of the food is obtained on the ground. The ♂ gives its song from the top of a bush or from wires. Outside the breeding season it is gregarious, and usually frequents agricultural land, stackyards, etc.

The nest is usually on the ground in grass or a tuft of herbage, but sometimes it is sited in thick bushes or hedges up to 2 m from the ground. The nest is a loose cup of grasses, lined with fine grass, roots and hair. It is built by the ♀ who also incubates alone. The young are fed by the ♀ though the ♂ may assist. The ♂♂ are often polygamous and are recorded as having as many as seven ♀♀.

The species is mainly resident and sedentary. In winter it often forms large flocks, sometimes in association with Larks, and may wander locally. In parts of the range, especially northern and western areas in Britain and Ireland, it has suffered a decline in population for reasons which are not clear. the species has generally benefited from the increase in agriculture in Europe, which has created the open landscapes it favours.

Length:	18 cm
Wing length:	8.9–10.1 cm
Weight:	*ca* 50 g
Voice:	Song: rapid jangling noise like rattling of bunch of keys. Call: 'quit' and 'quit-it-it'
Breeding period:	Mid-April to July. 2 (–3) broods per year
Size of clutch:	4–5 (1–7) eggs
Colour of eggs:	White ground, buffish or grey-blue, with large blotches, spirals and scribbles
Size of eggs:	23.3×18 mm
Incubation:	12–14 days, beg. before clutch complete
Fledging period:	Nidicolous; leaving nest at 9–12 days, half-fledged

nmark (Gé)

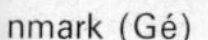

t) Singing ♂, Macedonia, Yugoslavia, 1969 (Li)

Macedonia, Yugoslavia, 1969 (Li)

Black-headed Bunting

(Emberiza melanocephala)

The Black-headed Bunting inhabits warm, open country with scattered bushes, cultivated land and steppe grassland as well as gardens, orchards and lightly wooded areas or *maquis*.

It feeds on seeds, grain fruits and a variety of insects including grasshoppers and earwigs. It feeds mostly on the ground, though it perches readily on bushes, posts and wires. Outside the breeding season it is gregarious and often forms large flocks. The ♂ gives its song from an exposed perch such as a tree-top or from wires. It sometimes flies from the song perch with legs dangling, like a Corn Bunting.

The nest-site is on or near the ground usually in thick cover, though sometimes 2–3 m up in a small tree. The nest is a cup of grasses, plant stems and other plant material lined with fine grass. It is built by the ♀ who also incubates alone. She feeds the young though sometimes the ♂ may assist.

The species is entirely migratory, wintering mainly in north-west India. It is prone to overshooting the breeding area on spring migration and occurs as an accidental visitor in western Europe, including Britain.

Migration: Departs breeding areas in late August to October, returning in April–May.

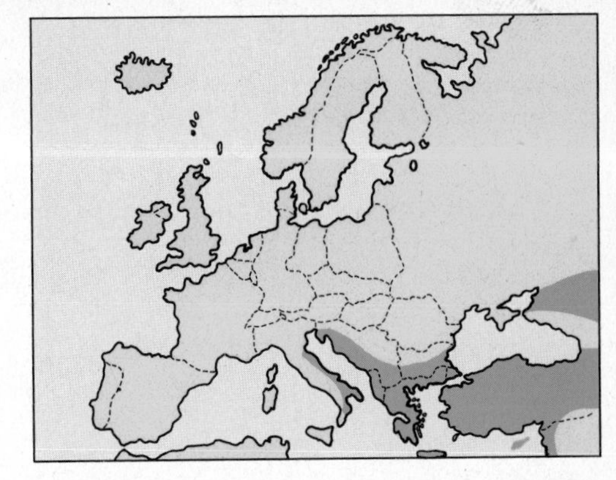

Length:	17 cm
Wing length:	8.4–9.8 cm
Weight:	*ca* 35 g
Voice:	Loud 'tsitt' and soft 'tseeh'; also gentle 'chup'
Breeding period:	Mid-May to June. 1 brood per year. Replacement clutch possible
Size of clutch:	4–5 (–7) eggs
Colour of eggs:	A few spots on a light greenish to bluish ground, concentrated at the blunt pole
Size of eggs:	22.5×16 mm
Incubation:	*ca* 14 days, beg. before clutch complete
Fledging period:	Nidicolous; leaving the nest half-fledged

. ♀, Macedonia, Yugoslavia, 1969 (Li)

Nearly-fledged nestlings, Greece, 5.6.1979 (Li)

Greece, 10.5.1972 (Li)

ʼt) ♂ singing, Macedonia, Yugoslavia, 1969 (Li)

Yellowhammer (Emberiza citrinella)

The Yellowhammer forms a sibling species with the Pine Bunting and sometimes interbreeds with that species (see p. 319). It occurs in light, open woodland, heathland, cultivated areas, waste ground, roadside hedges and bushy scrub. In winter it is common on stubble fields and other agricultural land. It feeds mainly on grain, seeds of grasses and weeds, and some fruits such as blackberry and a variety of small insects, spiders, earthworms, etc. Most of the food is obtained on the ground, though the bird perches freely on bushes, trees and wires. In winter it is gregarious, and often associates with other bunting and finches in mixed flocks.

The nest-site is usually on the ground, well concealed in a tuft of vegetation or at the edge of bushes or a hedgerow. Sometimes it will nest a little above the ground in a thick bush. The nest is a cup of plant stems, grass and other plant material, lined with hair or fine grass. It is built by the ♀ who also incubates alone, fed by the ♂. Both parents tend the young.

The Yellowhammer is mainly resident, though local dispersal takes place in autumn and winter. Birds from the extreme north of the range are migratory and winter in western Europe, Mediterranean countries and Asia Minor.

Migration: Autumn movements mainly late September to November, with return passage in March–May.

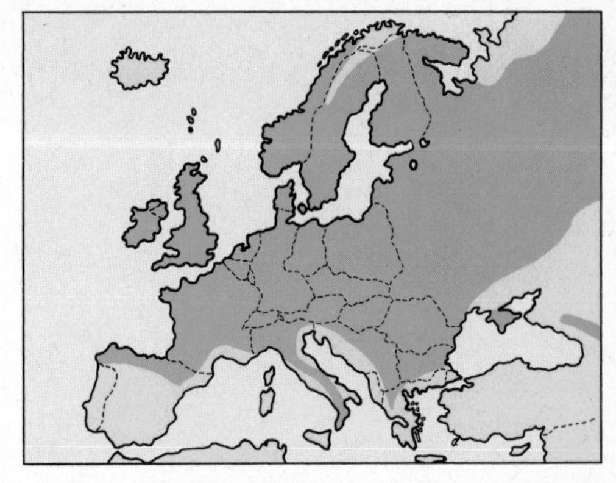

Length:	17 cm
Wing length:	8.0–9.3 cm
Weight:	*ca* 30 g
Voice:	Metallic 'tink' or 'twick'; also 'tillip'. Song: 'little bit of bread and no cheese'
Breeding period:	End of April to August. 2 (–3) broods per year
Size of clutch:	3–5 (2–6) eggs
Colour of eggs:	Basic colour white, reddish grey or bluish, with dark hairlines and scrawls, esp. at blunt pole
Size of eggs:	22×16 mm
Incubation:	12–14 days, beg. when clutch complete
Fledging period:	Nidicolous; leaving nest at 9–14 days, half-fledged

t nest, Bavaria (Diep)

Fledged young bird, Bavaria, July 1979 (Pf)

Bavaria, 6.5.1979 (Pf)

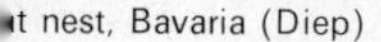

) Ad. ♂, Lower Saxony, July 1978 (Di)

Cirl Bunting (Emberiza cirlus)

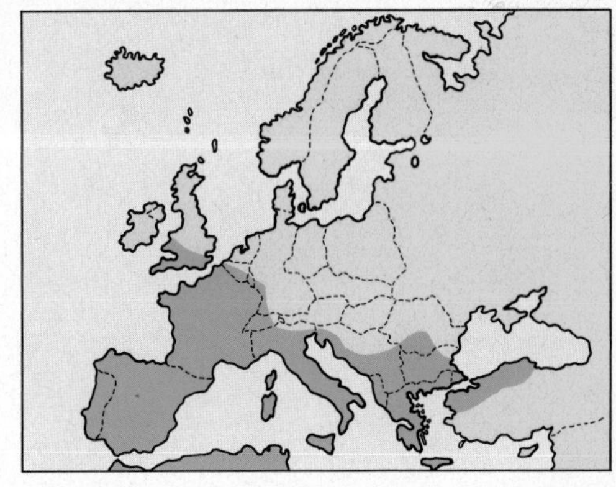

The Cirl Bunting is found in warm, sunny areas with scattered trees and bushes, woodland edges, parkland and cultivated land with scattered trees and roadside hedges as well as scrubby vegetation in hilly country.
It feeds on grain and seeds of grasses and weeds, berries, insects and larvae including grasshoppers and Lepidoptera. It feeds mainly on the ground, though it perches readily in tall trees or bushes. The song is often given from a tall tree or from a hedge or wires. It is gregarious in winter and often forms small parties, sometimes with Yellowhammers or other buntings.
The usual nest-site is in thick bushes, low trees or hedgerows, and is well concealed. Sometimes it will nest on the ground in a tuft of vegetation. The nest is a cup of roots, plant stems and grasses, lined with fine grass and hair. It is built by the ♀ who also incubates alone, and feeds the young, though the ♂ may assist with the latter.
The species is mainly resident, though prone to wandering in winter. The breeding range in Britain has contracted greatly in the present century, probably due to climatic changes and to the effect of hard winters.

Length:	16.5 cm
Wing length:	7.1–8.2 cm
Weight:	*ca* 25 g
Voice:	Drawn-out 'tseep'. Flight call: 'sissi-sissi-sip'. Song: monotonous, similar to Lesser Whitethroat
Breeding period:	End of April to August. 2–3 broods per year
Size of clutch:	3–4 (2–5) eggs
Colour of eggs:	Grey-white to reddish-brown ground, with blackish-brown spots, hairlines and scrawls
Size of eggs:	21×16 mm
Incubation:	11–12 days, beg. from complete clutch
Fledging period:	Nidicolous; leaving nest at 11–14 days

Greece, 7.7.1977 (Schu)

Greece, 5.6.1976 (Li)

eft) ♂ at nest, Greece, 10.6.1976 (Li)

Ortolan Bunting (Emberiza hortulana)

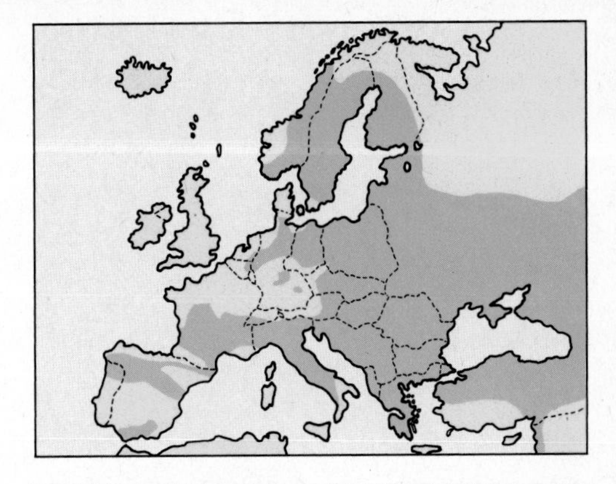

The Ortolan Bunting is found in dry, open country with sparse vegetation, often in arable land, open *maquis* scrub, thickets and woodland edges. It occurs in both lowland and dry mountain slopes, often on sandy or rocky ground.
It feeds on grain and small seeds as well as insects, including grasshoppers, locusts, beetles, etc. The young are fed mainly on insects or larvae. It feeds mostly on the ground and is a rather shy and secretive bird. On migration and in winter it is gregarious and often forms flocks with Tree Pipits on migration. It perches in bushes, trees and on wires and the ♂ gives his song from such a perch.
The nest is placed on the ground in the cover of thick herbage or under a bush. It is a cup of grasses and roots with a lining of finer plant material or a few feathers. The nest is built by the ♀ who also incubates alone. Both parents feed the young.
The Ortolan Bunting is entirely migratory, and winters in tropical Africa south of the Sahara, mainly in the north-east of the continent. It occurs regularly in Britain, mainly on the east coast as a passage-migrant in small numbers.
Migration: Departs breeding areas in August–September, returning in April–May, and into June in the north.

Length:	16 cm
Wing length:	8.0–9.2 cm
Weight:	*ca* 25 g
Voice:	Loud 'tsip' and 'pwit' or 'tseu'. Song resembles Yellowhammer's
Breeding period:	Beginning of May to June. Usually double-brooded. Replacement clutch possible
Size of clutch:	4–6 (–7) eggs
Colour of eggs:	Ash-grey to dirty pink, with sparse dark patches, and overlaid with squiggles
Size of eggs:	20 × 15.5 mm
Incubation:	11–12 days, beg. after clutch complete
Fledging period:	Nidicolous; leaving nest at 9–10 days, unable to fly for some days

Turkey, 4.6.1973 (Li)

Fledged young bird, Turkey, 9.6.1977 (Schu)

North Rhine–Westphalia, 18.5.1966 (Sieb)

t) ♀, Turkey, 28.6.1977 (Li)

Cretzschmar's Bunting

(Emberiza caesia)

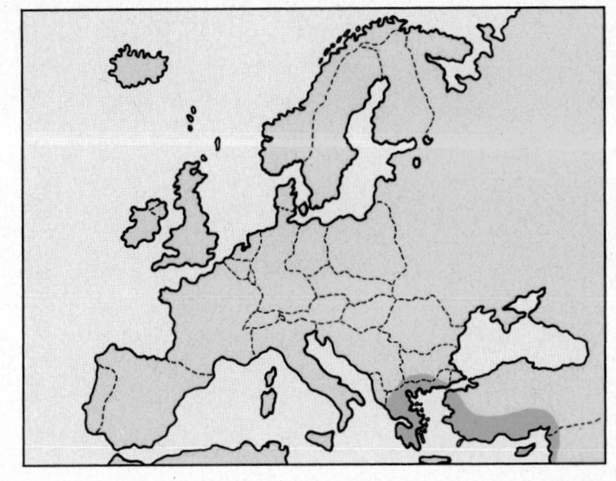

The Cretzschmar's Bunting is very similar in appearance and habits to the Ortolan Bunting and in some areas the range and habitat is shared. It occurs in rather arid areas such as dry hill-slopes with scrub and rocky areas, open *maquis* scrub and grassy areas with little cover. In mountainous areas it tends to occur at a lower altitude than the Ortolan.
The diet is probably very similar to that of the Ortolan.
The nest is placed on the ground in thick herbage or below a bush. It is a cup of plant stems and roots, lined with finer plant material. Little is known of the breeding biology, though it is probably much like that of the Ortolan.
The species is migratory, wintering in north-east Africa, mainly coastal Eritrea and the Sudan.
Migration: Leaves breeding areas in late August to September, returning in late March to April.

Length:	16 cm
Wing length:	*ca* 8.5 cm
Weight:	*ca* 25 g
Voice:	Insistent, harsh -styip'. Song refrains shorter than Ortolan's
Breeding period:	End of April to mid-June. 2 broods per year
Size of clutch:	4–5 (–6) eggs
Colour of eggs:	Base colour darker than Ortolan's, with more lines and squiggles
Size of eggs:	19.5×15 mm
Incubation:	11–12 days, beg. from complete clutch
Fledging period:	Nidicolous; details unknown

The **Cinereous** or **Ashy-headed Bunting** (*Emberiza cineracea*) breeds in Asia Minor and on some of the Greek islands. It occurs in rather dry, rocky areas with scattered scrub vegetation, often in mountainous areas. Very little is known of its diet, habits or breeding biology though it probably resembles the Cretzschmar's Bunting.
It is a migrant, wintering in the same areas in Africa as the Cretzschmar's Bunting.

♀, Lesbos, Greece, 15.5.1974 (Fe)

Young birds, Lesbos, Greece, 15.5.1974 (Fe)

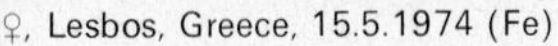

t) Ad. ♂, Lesbos, Greece, 16.5.1974 (Fe)

Rock Bunting (Emberiza cia)

The Rock Bunting is found in rocky and mountainous areas, often with little or no cover, and in areas with a growth of scrub or trees. In some areas it inhabits vineyards and cultivated ground down to sea-level.
The diet includes grain and seeds, small insects and other invertebrates. It feeds principally on the ground, perching on rocks and stones, though it will perch in trees where present. It is not gregarious as a rule but sometimes flocks in winter and may associate with other bunting, notably Cirl Bunting.
The usual nest-site is in a hole among rocks or in a wall; sometimes it will nest in low bushes among stones. The nest is a cup of grasses, plant stems, roots and moss, lined with roots and hair. It is built by the ♀ who also incubates alone. Both parents tend the young, which are fed largely on insects.
The Rock Bunting is a resident bird, though it moves to a lower altitude in winter and sometimes wanders outside the breeding range. Birds from the northernmost part of the range tend to disperse southwards in autumn.
The species has occurred as a vagrant in Britain and other north European countries.

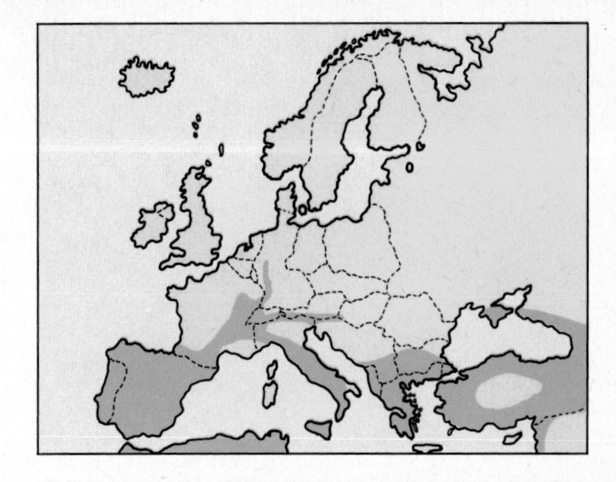

Length:	16 cm
Wing length:	8.4 cm
Weight:	*ca* 25 g
Voice:	Call: 'seea', also 'tzit'. Song: 'zi-zi-zi-zeer', resembling Reed Bunting
Breeding period:	April–July. 1, sometimes 2, broods per year
Size of clutch:	4–5 (–6) eggs
Colour of eggs:	Basic dirty white to brownish, thickly overlaid with dark tangled hairlines
Size of eggs:	21×16 mm
Incubation:	12–13 days, beg. from complete clutch
Fledging period:	Nidicolous; leaving nest at 10–13 days

♀, Carinthia, Austria, 4.7.1975 (Zm)

Ad. ♂, Carinthia, 4.7.1975 (Zm)

Turkey, 7.6.1977 (Li)

) Ad. ♂, Lower Franconia, West Germany, 3.6.1978 (Sch)

Reed Bunting (Emberiza schoeniclus)

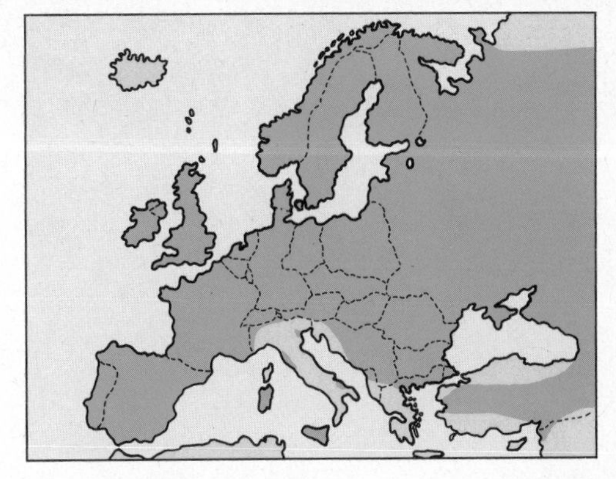

The Reed Bunting occurs in a variety of marshy or waterside habitats including reed-beds, fresh and saline marshes with rank vegetation, bushes and scrub bordering lakes or river, along ditches, swampy clearings in woodland and into shrub tundra in the north. It feeds on seeds of marsh-plants and grasses, grain, small snails and crustaceans, and a variety of insects and their larvae. Much of the food is obtained among reeds and bushes where it perches freely, often clinging to reed-stems or other plants. It will make short flights to catch insects on the wing. It also feeds on the ground and frequents cultivated land on migration and in winter, often in company with other buntings and finches.

The usual nest-site is on the ground or just above it, either in a tuft of vegetation or in a low bush. The nest is a cup of grass, plant stems, leaves and moss, lined with flower-heads and fine grass. It is built by the ♀ who also takes the major role in incubation, though the ♂ may assist. Both parents tend the young.

The Reed Bunting is mainly migratory. The northern and eastern populations migrate to winter in western or southern Europe. Birds from the rest of the range tend to leave the breeding areas and winter on coastal marshes, agricultural land and lowland areas.

Migration: Autumn movements mainly September–October, returning in late March–May.

Length:	15 cm
Wing length:	7.0–8.1 cm
Weight:	20 g
Voice:	Drawn-out 'tseeh'; also 'ching'. Song: 'tweek-tweek-tweek-titick'
Breeding period:	End of April to July. 2 broods per year, sometimes 3
Size of clutch:	5–6 (4–7) eggs
Colour of eggs:	Olive-grey to grey-green with large black markings, hairlines and squiggles
Size of eggs:	20×14.8 mm
Incubation:	13–14 days, beg. from complete clutch
Fledging period:	Nidicolous; leaving nest at 10–13 days

♀, Seewinkel, Austria, 1970 (Li)

Nestlings, Bavaria, 10.5.1972 (Pf)

Bavaria, 2.5.1974 (Pf)

) Ad. ♂, Seewinkel, Austria, 1970 (Li)

Little Bunting (Emberiza pusilla)

The Little Bunting breeds in birch and willow scrub in the tundra and in sub-arctic birch woods. It is also found in willows beside rivers in the northern taiga zone and low-lying swamps.
It feeds mostly on small insects and larvae during the breeding season, and on small seeds of grasses and other low plants during the rest of the year.
It spends most of the time on the ground or in low bushes and similar vegetation.
The nest-site is on the ground among tree-stumps, moss or low cover and is usually concealed in a tuft of grass or moss. The nest is a cup of grasses, moss and leaves, lined with fine grass and hair. Both sexes share incubation and the care of the young.
The Little Bunting is entirely migratory, wintering in India and south-east Asia, where it occurs in scrub and cultivated land. It occurs as an accidental visitor to western Europe, including Britain, and a few may overwinter in Europe.
Migration: Departs breeding areas in September–October, returning in April–May and early June.

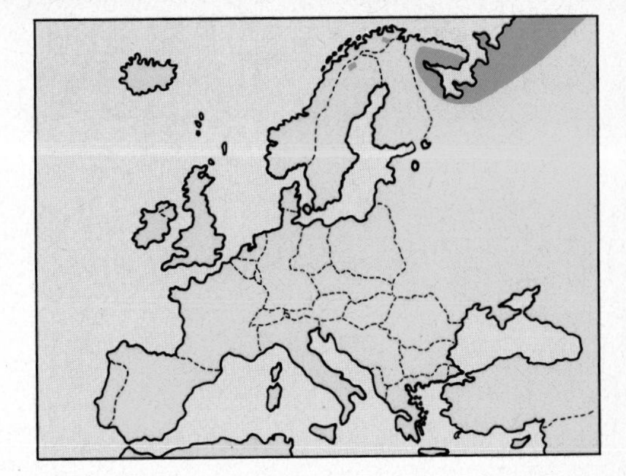

Length:	13.5 cm
Wing length:	6.4–7.5 cm
Weight:	17 g
Voice:	Call: 'tic' or 'pwit'. Song: rather Robin-like twittering
Breeding period:	Beginning of June to August. 1 (2) broods per year. Replacement clutch possible
Size of clutch:	4–5 (3–6) eggs
Colour of eggs:	Light grey to greenish ground, with sparse patches and lines, esp. at blunt pole. Very variable
Size of eggs:	18.5×14 mm
Incubation:	11–12 days, beg. from complete clutch
Fledging period:	Nidicolous; leaving nest as early as 6–8 days, able to fly after a further 3 days

. ♂, north-east Finland, 23.6.1976 (Gr)

Reed Bunting, winter plumage, Småland, Sweden, 12.1.1963 (Chr)

North-east Finland, 23.6.1976 (Gr)

t) Brooding ♀, north-east Finland, 23.6.1976 (Gr)

Rustic Bunting (Emberiza rustica)

The Rustic Bunting inhabits marshy areas with birch, willow or poplar trees on the edge of the taiga forests. It is also found in wet heathland with scrub and scattered trees and in dense spruce-forests with a groundcover of thick moss.

It feeds mainly on small seeds and other plant matter, the young being fed on seeds. In winter-quarters it takes rice and grain. It is largely terrestrial, hopping about among moss, tree-stumps and other low cover. It will perch in trees and bushes and the ♂ usually sings from such a perch.

The nest is on the ground in cover of grass or moss or in the lower branches of a bush. It is constructed of grass, plant stems and moss, lined with fine grass, roots and hair. Incubation is by the ♀ alone, though both parents feed the young.

The Rustic Bunting is entirely migratory, wintering in the temperate regions of east Asia and to a lesser extent in Turkestan. The species has extended its range to the west in the last century and occurs in west European countries and is an accidental visitor on both spring and autumn passage.

Migration: Departs breeding areas in September–October, returning in April–May and early June.

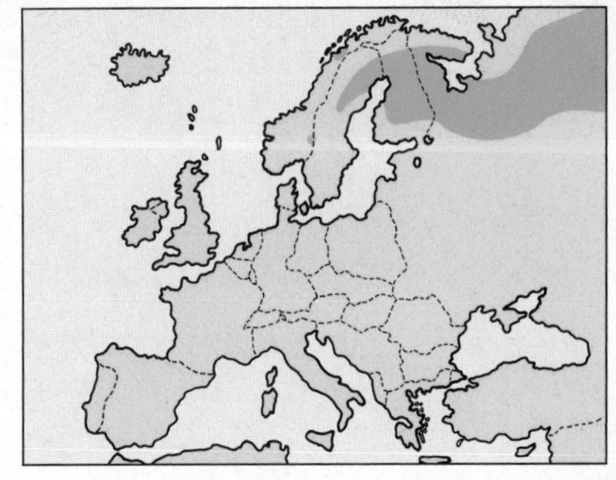

Length:	14.5 cm
Wing length:	7.2–8.1 cm
Weight:	18 g
Voice:	Call: short, hard 'tic' like Robin. Song: like Robin with short phrases
Breeding period:	End of May to July. 1, rarely 2, broods per year. Replacement clutch possible
Size of clutch:	4–5 (6) eggs
Colour of eggs:	Dirty white to light green, with thick brown or olive patches
Size of eggs:	20×15 mm
Incubation:	12–13 days, beg. from complete clutch
Fledging period:	Nidicolous; leaving nest at *ca* 14 days

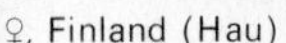

♀, Finland (Hau)

♀ at nest, central Finland, 14.6.1979 (Gr)

) Ad. ♂, central Finland, 14.6.1979 (Gr)

Yellow-breasted Bunting

(Emberiza aureola)

The Yellow-breasted Bunting breeds in the willow and birch zones of northern Europe. it occurs along river banks and inundation zones and in forest clearings. In some areas it inhabits dry scrub in open grassy country. The diet is mainly insects and larvae. The species also takes small seeds of grasses and other plants and in the winter-quarters, rice and grain.
The usual nest-site is in a low bush or amongst tall herbage. The nest is a cup of plant stems and grass lined with fine grass. Both sexes take part in nest-building. The ♀ incubates alone and both parents tend the young.
The species is entirely migratory, wintering in tropical south-east Asia. It occurs as an accidental visitor to western Europe, mainly on autumn passage.
Migration: Departs breeding areas in late August to September, returning in April–May or early June.

Pine Bunting (Emberiza leucocephalos) This species is closely related to the Yellowhammer (see p. 302). It inhabits coniferous and birch woodland, open country near water and wooded steppes. The behaviour, breeding biology and diet are very similar to those of the Yellowhammer.
The Pine Bunting is migratory, wintering in the temperate regions of the Middle East and Asia from Iran to China.
Migration: Departs breeding areas in September–November, returning in April–May. It is a rare vagrant in western Europe.

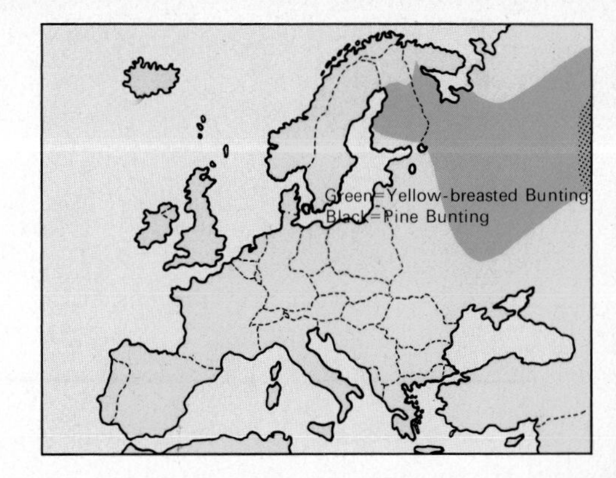

Length:	14 cm Yellow-breasted Bunting
Wing length:	7.1–8.0 cm
Weight:	18 g
Voice:	Call 'tic' or 'tsip'. Song: like Ortolan's, though higher and more rapid
Breeding period:	Beginning of June to August. 1 (2) broods per year. Replacement clutch possible
Size of clutch:	4–5 (6) eggs
Colour of eggs:	Light green to olive-green, with variable brownish smudging and a few dark spots
Size of eggs:	20.5×15 mm
Incubation:	*ca* 13 days
Fledging period:	Nidicolous; leaving nest at 13–15 days, independent at *ca* 28 days

Length:	17 cm Pine Bunting
Wing length:	8.2–10 cm
Weight:	30 g
Voice:	Call: 'tit' or 'tit-it-up'. Song: rather Robin-like
Breeding period:	End of May to beginning of August. 2 broods per year
Size of clutch:	4–5 (–6) eggs
Colour of eggs:	Whitish ground with dark grey or red-brown patches, scribbles and hairlines
Size of eggs:	21.5×16 mm
Incubation:	12–14 days, beg. from complete clutch
Fledging period:	Nidicolous; fledged at *ca* 2 weeks

ne Bunting ♀ (Zm)

Pine Bunting's clutch, breeding reservation (Pf)

ft) Yellow-breasted Bunting ♂ (Hau)

Snow Bunting (Plectrophenax nivalis)

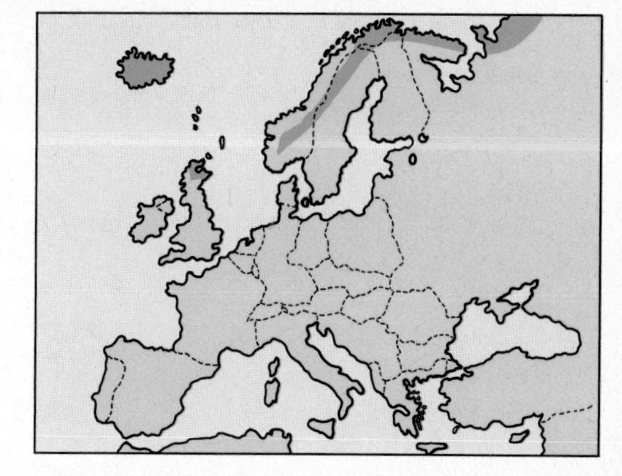

The Snow Bunting inhabits bare rocky tundra, scree slopes, mountain-tops and similarly barren areas. It is a town bird in southern Greenland and is more coastal in distribution than the Lapland Bunting.
The diet is mainly insects in summer, especially midges and larvae of small Lepidoptera, which is the main diet of the young. It also takes buds and small shoots and a variety of seeds, the latter being important in winter. It also takes small crustaceans, sandhoppers and shore-loving insects in winter. It is essentially a ground-living bird, though it will perch on wires and corn-stacks, etc.
The nest is on the ground, usually concealed in a rock-crevice, or among stones or driftwood. It also nests in holes in walls or discarded tins, pipes, etc. near human habitation. The nest is a cup of moss and lichens with fine grass, hair and feathers for a lining. It is built by the ♀, often with the ♂ in attendance. The ♀ incubates alone, fed by the ♂ who also brings food for the young in the first few days after hatching. Later both parents bring food. The ♂♂ are sometimes polygamous.
The Snow Bunting is partly migratory, with birds wintering, often in large flocks, on coastal areas or open country in north-west Europe, including Britain. Some birds are resident, including much of the Icelandic population.
Migration: Departs breeding areas in September–November, returning in April–May.

Length:	16.5 cm
Wing length:	10–11 cm
Weight:	31 g
Voice:	Calls: 'See-oo' and liquid 'tirr-irr-irr-rip'. Song: high musical, rather Lark-like
Breeding period:	End of May to July. 1, rarely 2, broods per year. Replacement clutch possible
Size of clutch:	5–6 (4–8) eggs
Colour of eggs:	Buffish grey to light blue with rusty or dark-brown patches. Colour and markings very variable
Size of eggs:	22×16 mm
Incubation:	10–15 days, beg. from complete clutch
Fledging period:	Nidicolous; leaving nest at 10–14 days

lardanger vidda, Norway, 25.6.1969 (Pl)

Juv., Norway, 5.7.1977 (Di)

Dovrefjell, Norway, 25.7.1977 (Di)

) Ad. ♂, Dovrefjell, Norway, 5.7.1977 (Di)

Lapland Bunting (Calcarius lapponicus)

The Lapland Bunting is found in open shrub tundra with willow, birch and other scrub as well as on mountain-tops and swampy moss tundra. In winter it frequents coastal areas such as salt-marshes, meadows and similar flat areas.

It feeds on insects and larvae, mainly midges, in summer, and this is the main diet of the young. It also takes small seeds, buds and shoots. In winter it will feed on grain and other large seeds. It is mainly terrestrial, but sometimes perches on bushes in the winter-quarters, when it will mix with other buntings and larks.

The nest-site is on the ground, usually in a hollow or on a bank or hummock. The nest is a cup of grasses, roots and moss, lined with fine grass, hair or feathers. The nest is built by the ♀, who also does most of the incubation, though the ♂ will assist at times. Both parents tend the young. A few pairs have bred in Britain recently.

The Lapland Bunting is migratory, wintering in the coastal lowlands of north-west Europe. It is gregarious in winter and on migration.

Migration: Departs breeding areas in late August to October, returning in March to early May.

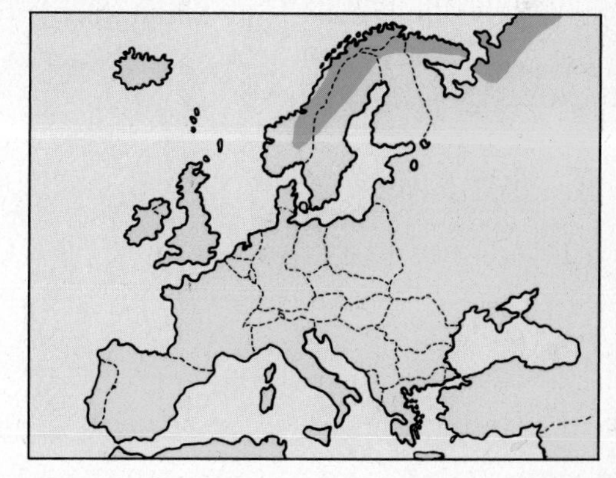

Length:	15 cm
Wing length:	8.3–9.6 cm
Weight:	24 g
Voice:	Call: rattling 'tick it tick-teu'. Song: repeated 'tee-too-ree', somewhat Lark-like
Breeding period:	Mid-May to Mid-July. 1 brood per year. Replacement clutch possible
Size of clutch:	4–6 (2–7) eggs
Colour of eggs:	Very variable. Dirty brown with thick spots and mottling
Size of eggs:	20.5 × 15 mm
Incubation:	10–14 days, beg. with last egg
Fledging period:	Nidicolous; leaving nest at 8–10 days, able to fly after further 4–5 days

rooding, Fokstumyra, Norway, 25.6.1968 (Sy)

Almost-fledged nestlings, Norway, 28.6.1968 (Sy)

Fokstumyra, Norway, 24.6.1972 (Sy)

) Ad. ♂, Fokstua, Norway, 19.6.1969 (Pl)

Chaffinch (Fringilla coelebs)

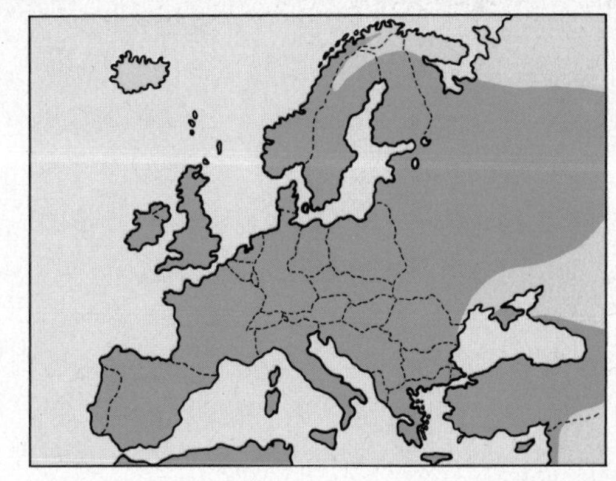

The Chaffinch is found in broad-leaved, mixed and coniferous woodlands, parkland, gardens, hedgerows, copses and thickets. It occurs from lowland areas to upland and montane zones. It is common in cultivated areas as well as in pure spruce-forest and Mediterranean pines. In winter it will frequent farmland and other open country. The diet is mainly vegetable matter such as seeds, grain, fruit, buds and shoots. Insects are also taken, as well as spiders, earthworms and snail eggs.

During the breeding season much of the insect food is obtained in the tops of trees and bushes. During the rest of the year it is much more terrestrial, though it perches freely on bushes and trees. It is strictly territorial when nesting but gregarious in winter, and often mixes with other finches, especially Brambling.

The nest-site is normally built into a fork in a tree or tall bush. It is a neat cup of moss, grasses, roots and feathers bound with spiders' webs and decorated with lichen and bark fragments which serve as camouflage. The nest is built by the ♀, who also incubates alone, though the ♂ has occasionally been recorded at the nest. Both parents tend the young.

The Chaffinch is a partial migrant. Birds from the north and east of the range winter in west and southern Europe and parts of the Middle East. Western birds, including the British population, do not normally move far from the breeding areas, though they wander over agricultural land in search of food.

Migration: Main movements in mid-September to November and in March–May.

Length:	15 cm
Wing length:	7.8–9.0 cm
Weight:	20 g
Voice:	Loud 'pink-pink', also 'tsup' and 'wheet'. Song: loud and vigorous short phrases ending in 'choo-u-oo'
Breeding period:	End of April to July. 1–2 broods per year
Size of clutch:	4–5 (2–8) eggs
Colour of eggs:	Bluish ground, with pink clouding and a few purple-brown patches and smudges, esp. at blunt pole
Size of eggs:	20×15 mm
Incubation:	11–13 days, beg. from penultimate egg
Fledging period:	Nidicolous; leaving nest at 13–14 days

nd ♀, feeding young, Bavaria, 10.5.1978 (Pf)

Nestlings, 7 days old, Bavaria, 10.5.1978 (Pf)

Amperaue, Bavaria, 1973 (Li)

) ♂ at nest, Bavaria, 10.5.1978 (Pf)

Brambling (Fringilla montifringilla)

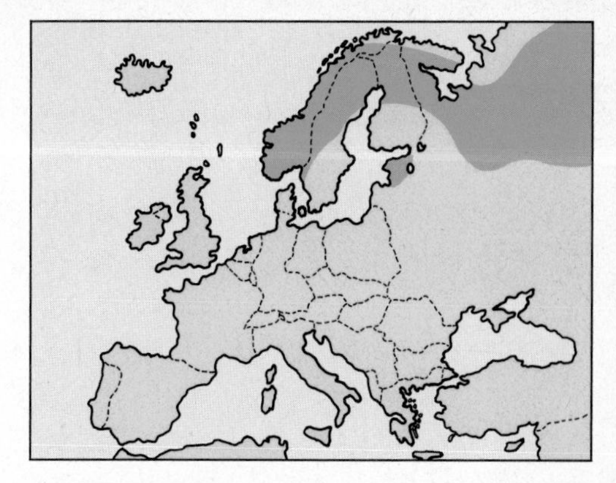

The Brambling is found in dry, taiga forests, birch and willow woods, forest edges and sub-arctic birch scrub. It is the northern ecological representative of the Chaffinch. In winter it inhabits beechwoods, plantations and open farmland, where it mixes with other finches and buntings, especially with Chaffinches.
It feeds mainly on insects in the breeding season, including many caterpillars, which are the main food of the young. It also takes seeds and berries, beechmast and hornbeam seeds in winter. It is highly gregarious on migration and in winter.
The nest-site is normally in the fork of a tree 1–10 m from the ground, though sometimes it will nest in bushes. The nest is a neat cup, rather like that of the Chaffinch, but larger. It is constructed of moss, grass and hair with a binding of spiders' webs and decorated with fragments of birch bark and lichen. Nest-building is done by the ♀, who also incubates alone. Both parents tend the young.
The Brambling is entirely migratory, wintering in the temperate areas of Europe and across Asia to Japan. Scandinavian birds largely winter in western Europe, including Britain. Very large flocks occur where food is abundant.
Migration: Departs breeding areas in September–October, returning in March–May, usually arriving on breeding grounds about a month after the Chaffinch. The species has occasionally nested in Britain.

Length:	15 cm
Wing length:	8.3–9.3 cm
Weight:	23 g
Voice:	Flight call: 'chucc-chucc-chucc' and 'tsweek'. Song: sweet and melodious similar to Redwing
Breeding period:	May–July. 1 brood per year. Replacement clutch possible
Size of clutch:	5–7 (4–9) eggs
Colour of eggs:	Like Chaffinch's, with more numerous spots and scrawls
Size of eggs:	19.5 × 14.5 mm
Incubation:	11–12 days
Fledging period:	Nidicolous; leaving nest at *ca* 11 days

n breeding plumage, Odenwald, West Germany, rch 1962 (Pf)
t) ♂ in non-breeding plumage (Li)

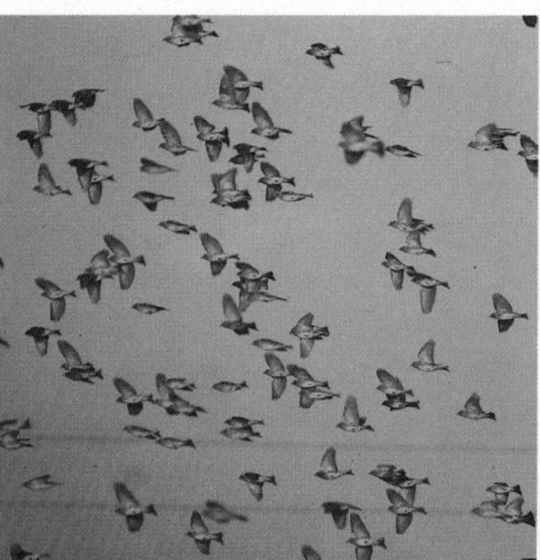

Skanör, Sweden, 14.10.1975 (Chr)

Southern Norway, 13.6.1965 (Sy)

Serin (Serinus serinus)

The Serin inhabits open woodland, gardens, parkland with scattered trees, vineyards, orchards and scrub with taller trees. It prefers warm, sunny areas such as forest-edges. In some areas it is common in towns and villages. In winter it joins with other finches in wandering groups.
The diet is almost entirely seeds of weeds and grasses, and of trees such as elder and birch. It also takes some buds and shoots.
The nest is sited in a tree or bush, often in a fork or towards the end of branches. It is a neat cup of plant stems, roots, moss and lichens, lined with feathers, hair and plant down. The nest is built by the ♀ who also incubates alone, though rarely the ♂ will assist. Both parents tend the young.
The Serin is migratory except in the southern part of the range. It winters in Mediterranean countries.
Migration: Autumn passage mainly October–November, returning in March–April. It occurs as an accidental in Britain and has bred on occasions.

Citril Finch (*Serinus citrinella*) This species is confined to the mountains of central and southern Europe, where it inhabits light coniferous woodland. it feeds on the seeds of conifers as well as seeds of plants. The nest is a delicate cup usually placed on the end of a branch. Incubation is by the ♀ with both parents feeding the young. The Citril Finch is largely resident, though it descends to lower altitudes in winter, when it often mixes with flocks of Redpolls and Siskins.

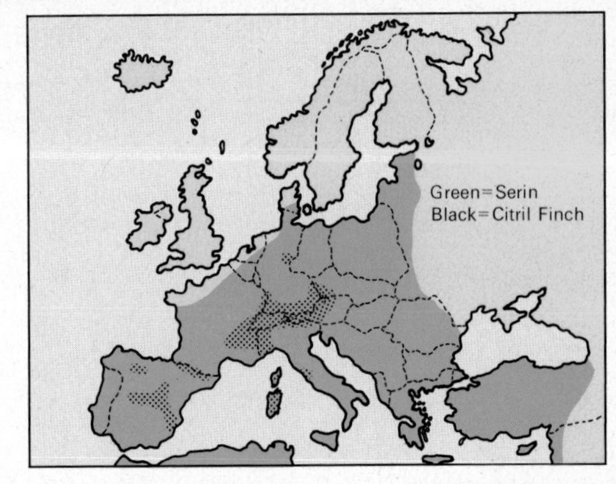

Length:	12 cm
Wing length:	6.4–7.4 cm
Weight:	12–13 g
Voice:	Rattling 'tirrililit'. Song: rapid jingle, sometimes with Canary-like trills
Breeding period:	Mid-March (southern Europe), otherwise end of April to July. 2 broods per year
Size of clutch:	4 (3–5) eggs
Colour of eggs:	Bluish white with red-brown and lilac patches, in a zone round the blunt pole
Size of eggs:	16.5×12.0 mm
Incubation:	13 days, beg. from complete clutch
Fledging period:	Nidicolous; leaving nest at 14 days

erin ♂, Hessen, 10.5.1970 (Qu)

eft) Serin ♀, feeding young, Hessen, 25.6.1977 (Tö)

Serin's clutch, Lower Saxony, 18.5.1976 (Sy)

Citril Finch, Upper Bavaria, 1970 (Schw)

Greenfinch (Carduelis chloris)

The Greenfinch is found in open habitat such as woodland edges, hedgerows, parks and gardens with tall trees, farmland with copses or scattered trees as well as scrub and thickets. In winter it frequents agricultural land, coastal grassland and similar open areas. It is gregarious in winter and often mixes with other finches in foraging parties. Though not really colonial, several pairs may nest in close proximity on occasion.
The diet is mainly seeds and grain, buds and shoots, beetles and other insects. The young are fed mainly on insects and pulped seeds. Much of the food is obtained on or near the ground. The ♂ has a circling, bat-like song-flight with slow wing-beats.
The nest-site is in a tree or bush, either in a fork or close up against the trunk. The nest is a rather untidy cup of plant stems, grass and moss, lined with fine plant matter, hair and a few feathers. The nest is built by the ♀, who also incubates alone. Both sexes feed the young, though the ♂ may take care of the fledglings whilst the ♀ begins a second clutch.
The Greenfinch is largely resident, though it wanders in winter, depending on the availability of food. Birds from the northern part of the range migrate to winter in temperate southern Europe and Mediterranean countries.
Migration: Passage takes place in October–November and March–April.

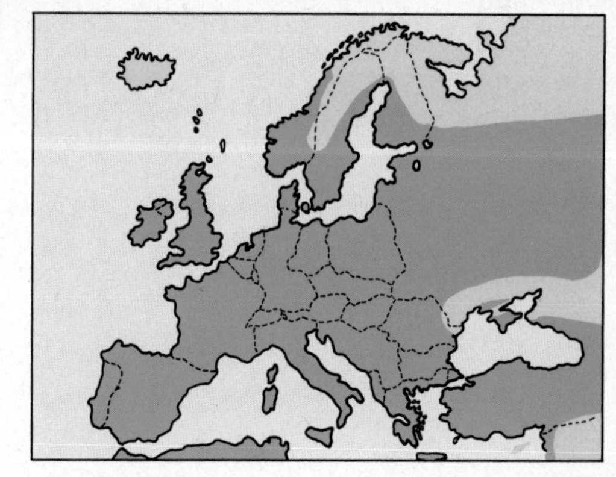

Length:	15 cm
Wing length:	8.3–9.0 cm
Weight:	*ca* 30 g
Voice:	'Chi-chi-chi chit' and long-drawn 'tsweee'. Song: twittering mixed with call notes
Breeding period:	End of April to September. 2–3 broods per year
Size of clutch:	4–6 (3–8) eggs
Colour of eggs:	Bluish white to buffish ground, with sparse rusty brown patches, esp. at blunt pole
Size of eggs:	20.3 × 14.8 mm
Incubation:	12–14 days, beg. before complete clutch
Fledging period:	Nidicolous; leaving nest at 13–14 days, not yet fully able to fly

nd ♀ at nest, Württemberg (Ca)

Nestlings, 2 days old, Upper Bavaria, 10.7.1979 (Pf)

Hansag, Austria, 1968 (Li)

t) ♂ sub-adult (Li)

Goldfinch (Carduelis carduelis)

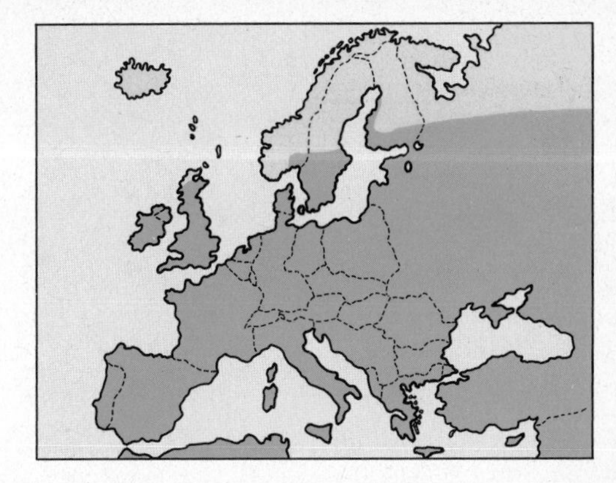

The Goldfinch inhabits areas similar to the Greenfinch, such as scattered trees in farmland, wasteland or gardens, woodland edges, orchards and woodland clearings. In winter it is common in farmland and scrub or waste areas with thistles and weeds.
The diet is mainly seeds, particularly thistle and other weeds, as well as birch and alder seeds. A variety of insects such as beetles, caterpillars and aphides are also taken. The species is very adept at taking the seeds from thistles, teasel and similar plants, and can climb up the stems and perform tit-like antics to get at seeds. The species is gregarious and is often seen in small parties. It does not usually mix with other finches in feeding flocks.
The nest is usually sited in a tree or tall shrub with open foliage. The nest is placed in twigs towards the end of a branch and is a neat cup of moss, lichen, wool, roots and plant down, lined with wool and plant down. It is built by the ♀, though sometimes the ♂ may bring materials. Incubation is by the ♀ alone, fed by the ♂. Both parents feed the young, who are dependent on them for a week or so after fledging.
The Goldfinch is migratory in eastern and northern Europe, wintering in Mediterranean countries and parts of the Middle East. The more western breeding birds, including those in Britain and Ireland, are mainly resident, though they disperse in winter.

Length:	13 cm
Wing length:	7.3–8.1 cm
Weight:	*ca* 16 g
Voice:	Call: 'stitellitt' and 'tswit-twit-twit'. Warning call: 'adee'. Song: liquid twitter, recalling Canary
Breeding period:	Beginning of April to July. 2–3 broods per year
Size of clutch:	5 (4–7) eggs
Colour of eggs:	Whitish-blue ground, with sparse red-brown spots, mostly at blunt pole
Size of eggs:	18 × 13 mm
Incubation:	12–14 days, beg. from complete clutch
Fledging period:	Nidicolous; leaving nest at *ca* 13–14 days

ɔldfinches seeking food (Li)

Nestlings, 8 days old, Upper Bavaria, 4.8.1977 (Pf)

Upper Bavaria, 29.7.1977 (Pf)

ft) Bavaria, 1968 (Li)

Redpoll (Carduelis flammea)

The Redpoll breeds mainly in larch or pine forests and in birch, willow and alder woods. In Britain plantations are favoured as well as birch and willow woods. It feeds mainly on seeds of birch, alder, larch and other conifers as well as on small insects. The young are fed largely on caterpillars. It feeds in a rather tit-like manner.

The nest is in a tree or bush and is a small, untidy cup of fine twigs, plant stems and grass, lined with plant down, hair and feathers. Often several pairs nest near each other. Incubation is by the ♀ alone, fed by the ♂. Both parents tend the young.

The northern populations are migratory, wintering in central and western Europe. Mass irruptions occur at times and the movements of the species are rather random.

Siskin (Carduelis spinus) The Siskin breeds mainly in light coniferous forest such as spruce, or in mixed woodland in both lowland and mountainous areas. It feeds on all kinds of conifer seeds as well as birch and alder. Some buds and insects are taken. The young are fed mainly on insects and pulped seeds. The nest is usually in a conifer, on the end of a branch. It is a small cup of fine twigs, moss, grass and wool with a lining of fine plant matter, hair and feathers. The ♀ incubates alone and both parents feed the young, though the ♀ feeds them with food brought by the ♀ at first.

Siskins are not really migratory but dispersive in winter. Sometimes quite large movements take place, and the species has reached the Mediterranean and North Africa.

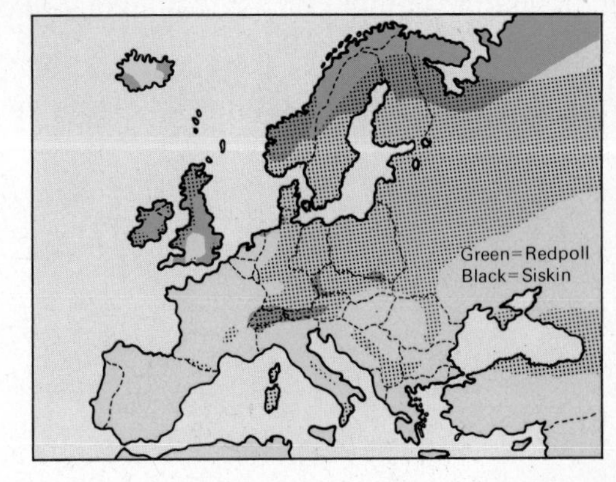

Length:	13 cm Redpoll
Wing length:	6.5–7.6 cm
Weight:	*ca* 14 g
Voice:	Metallic 'chutch-utch-utch', or trilling 'crrr' and 'tswit'. Song: combination of above sounds
Breeding period:	Begins at end April in south and lasts to mid-July in north. 1–2 broods per year
Size of clutch:	4–5 (3–8) eggs
Colour of eggs:	Bluish white to light blue-green with sparse spots, mostly in a crown at the blunt pole
Size of eggs:	16.9×12.5 mm
Incubation:	10–13 days, beg. from complete clutch
Fledging period:	Nidicolous; leaving nest at 11–14 days

Length:	12 cm Siskin
Wing length:	6.7–7.4 cm
Weight:	12–14 g
Voice:	Shrill 'tsuu' or 'tsyzing'. Flight call: 'tretteretet'. Song: rapid twitter
Breeding period:	April–June. 2 broods per year
Size of clutch:	4–5 (2–6) eggs
Colour of eggs:	Bluish white with a few rusty brown spots and twirls, esp. at blunt pole
Size of eggs:	16×12 mm
Incubation:	11–12 days, beg. from penultimate egg
Fledging period:	Nidicolous; leaving nest at 13–15 days

edpoll's clutch, Norway, 20.6.1968 (Sy)

Pair of Siskins, Carinthia, 2.2.1978 (Zm)

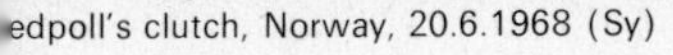

ɘft) Redpoll ♂ feeding ♀ at nest, Austria, 1979 (He)

Linnet (Carduelis cannabina)

The Linnet inhabits heathland, hedgerows, forest edges, plantations, cultivated land, shrub etc. It feeds mostly on seeds of low-growing plants and some insects, including caterpillars. The bird feeds mostly on or near the ground and often joins with Greenfinches and Tree Sparrows in winter to forage.
The nest-site is in a bush or in tall vegetation. The nest is a bulky cup of grass, plant stems, moss and small twigs, lined with hair and wool, sometimes with a few feathers. The ♀ incubates alone, fed by the ♀, who also brings food for the young at first. Later both birds bring food.
The species is a partial migrant; northern birds winter in the Mediterranean countries and the Middle East. The western and southern populations are largely resident, although some continental birds winter in Britain and some British birds move to Iberia or France.
Migration: Autumn movements mainly September–October, returning in March–April.

Twite (*Carduelis flavirostris*) The Twite is the upland counterpart of the Linnet, breeding on moorland, rough pasture and heathland. It nests on the ground in heather or other low cover and in bushes where available. In winter it moves to coastal areas, salt marshes and other low-lying areas. The species breeds in the high mountainous areas of Asia Minor and central Asia and is the only representative of the Tibetan faunal type in Europe.

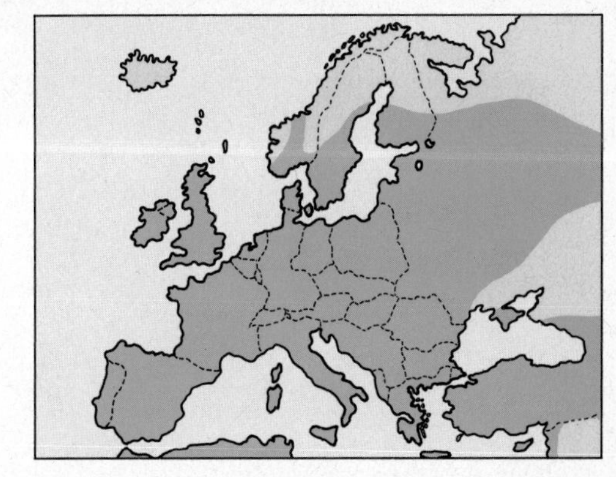

Length:	13 cm
Wing length:	7.5–8.2 cm
Weight:	*ca* 18 g
Voice:	Flight-call 'chit chit chit chit it'. Song: melodious with twittering and other notes
Breeding period:	End of April to August. 2–3 broods per year
Size of clutch:	5–6 (4–8) eggs
Colour of eggs:	Ground mostly light blue to dirty white, with red-brown spots and twirls, esp. at blunt pole
Size of eggs:	18×13 mm
Incubation:	11–12 days, beg. from complete clutch
Fledging period:	Nidicolous; leaving nest at 11–13 days

aria, July 1968 (Li)

Nestlings, 8 days old, Upper Bavaria, 12.7.1977 (Pf)

Upper Bavaria, 7.7.1977 (Pf)

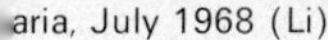

) ♂, Bavaria, July 1968 (Li)

Scarlet or Common Rosefinch

(Carpodacus erythrinus)

The Scarlet Rosefinch occurs in swampy woodland such as willow and alder thickets, forest edges, woodland near water meadows and marshes as well as bushes and thickets in cultivated areas. The species has greatly extended its range westwards since the 1930s.
It feeds almost entirely on seeds, buds and shoots, though some insects are taken, especially by young. It feeds in the manner of the Greenfinch, mainly on or near the ground, and can climb vertical plant stems to take seeds.
The nest is sited in a shrub or low tree, from ground-level to about 3 m. The nest is a rather loose, flat structure of plant stems and roots, lined with fine plant material and hair. It is built by the ♀, who also incubates alone. Both parents tend the young.
The species is entirely migratory, wintering in Asia from Iran to Indo-China. It inhabits the same swampy cover in winter and occurs in small parties. The species has become increasing regular on migration in western Europe, including Britain, both in autumn and spring passage. This may indicate that some of the population now migrates south-west and winters in southern Europe. Occasionally winter records in western Europe support this theory.
Migration: Departs breeding areas in late July to September, returning in April–May and early June in the north.

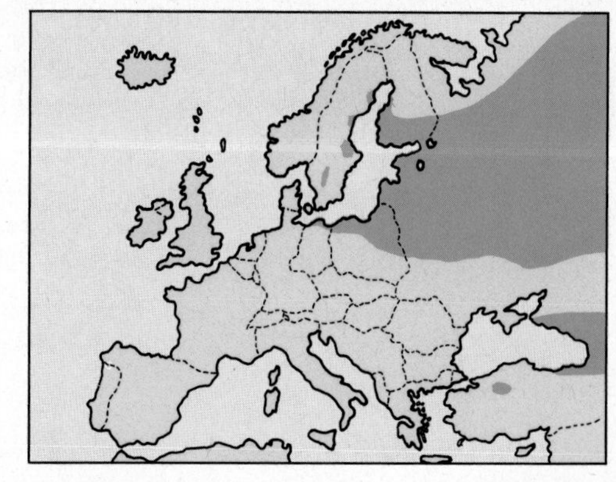

Length:	15 cm
Wing length:	7.8–8.6 cm
Weight:	*ca* 21 g
Voice:	Gentle 'hoo-eet' or 'twee-ekk'. Song: Oriole-like 'tiu-tiu-tiu'
Breeding period:	End of May to July. 1 brood per year
Size of clutch:	5 (3–6) eggs
Colour of eggs:	Blue-green with sparse dark brown patches, spots add wiggles at blunt pole
Size of eggs:	20×14.5 mm
Incubation:	11–12 days, beg. from complete clutch
Fledging period:	Nidicolous; leaving nest at 10–15 days

eeding young, Finland (Hau)

♂, Murnauer Marsh, Upper Bavaria, 8.6.1978 (Sieb)

Carinthia, 10.6.1977 (He)

t) ♂ at nest, Finland (Hau)

Pine Grosbeak (Pinicola enucleator)

This, the largest finch of the region, breeds in light taiga forest with larch, spruce or pines, northwards into the birch zone. In some areas it occurs in mixed forest with few conifers and in open juniper scrub.
It feeds on the seeds, buds and needles of conifers and the buds of birch and willow trees as well as the berries of elder and juniper and some insects. It feeds mainly in the tree-tops and is rather quiet and unobtrusive. It will also feed on the ground, particularly in the winter. It is often exceedingly tame and trusting. The nest is usually in a birch, conifer or juniper tree, rarely higher than 3 m from the ground. The nest is like that of a large Bullfinch, being a rather loose structure of twigs with an inner cup of fine plant material and moss. It is built by the ♀, who also incubates alone, being fed by the ♂. Both parents feed the young on regurgitated insects and plant matter.
The species is not truly migratory but is subject to dispersive movements which in some years become major irruptions. These movements depend upon availability of food and population levels. At times they result in birds occurring well outside the normal breeding range.
Migration: Irruptive movements usually occur in October–November and continue through the winter as food supplies become exhausted.

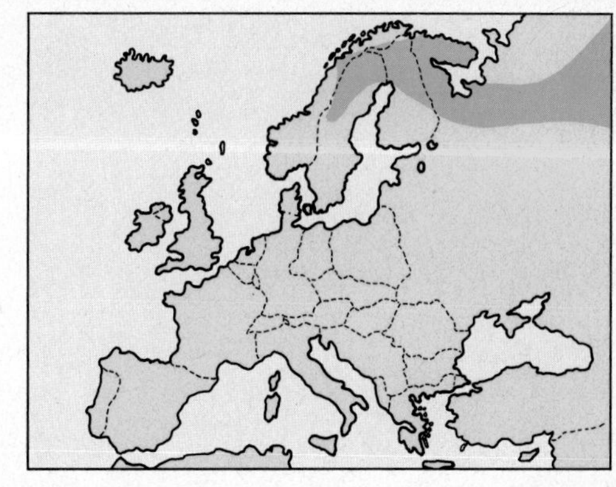

Length:	20 cm
Wing length:	10.5–11.5 cm
Weight:	60 g
Voice:	Call: trisyllabic 'tee-tee'tyoo'. Song: loud and piping with interspersed nasal sounds
Breeding period:	End of May to July. 1 brood per year. Replacement clutch possible
Size of clutch:	4 (3–5) eggs
Colour of eggs:	Blue-green with large dark brown and black spots, concentrated at blunt pole
Size of eggs:	26×18 mm
Incubation:	13–14 days, beg. from complete clutch
Fledging period:	Nidicolous; leaving nest at 14 days

ove and left) ♂, Finland (Hau)

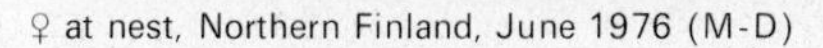

♀ at nest, Northern Finland, June 1976 (M-D)

Northern Finland, June 1976 (M-D)

Parrot Crossbill (Loxia pytopsittacus)

The Parrot Crossbill forms a sibling species with the Crossbill. It occurs in dry pine woods with mature trees and an understorey of heather. It will also occur in other coniferous forest with a high proportion of pines.
The diet is specialised and consists mainly of seeds of pine cones. Sometimes spruce and larch seeds are taken and in winter berries and catkins are important.
The nest is placed in a tall conifer, usually 3–14 m from the ground. It is a more substantial structure than that of the Crossbill, and is made of twigs with a cup formed of pine needles, moss, lichen and feathers. Incubation is by the ♀, who is fed by the ♂. He also brings food for the young at first but later both parents bring food. The young are fed on regurgitated pine seeds.
The species is not migratory but is subject to dispersal in autumn and winter, and sometimes mass irruptions take place. Often these movements coincide with those of the Crossbill, and the two species will form mixed flocks.

The **Scottish Crossbill** (*Loxia scotica*) has recently been recognised as a full species. It occurs in Caledonian pine-woods in northern Scotland and is largely resident. The size of the bill is intermediate between that of the Parrot and of the ordinary Crossbill. It feeds largely on seeds of Scots Pine.

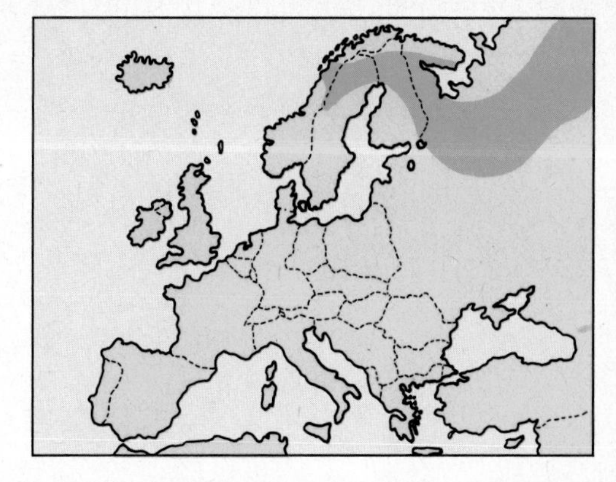

Length:	17.5 cm
Wing length:	9.8–11.0 cm
Weight:	52 g
Voice:	Call: a deep 'kop kop'. Song: resembles Crossbill's
Breeding period:	Irregular, depends on availability of food. December–August. 1 (2) broods per year
Size of clutch:	4 (3–5) eggs
Colour of eggs:	Dirty white or buffish ground, with sparse red-brown or dark-brown spots
Size of eggs:	23 × 17 mm
Incubation:	14–16 days
Fledging period:	Nidicolous; leaving nest at *ca* 25 days, but fed by the parents for some weeks longer

nest, Finland (Hau)

♀ at nest (Hau)

Clutch with newly hatched chick, Finland (Hau)

) Ad. ♂, Sweden (Ge)

Crossbill (Loxia curvirostra)

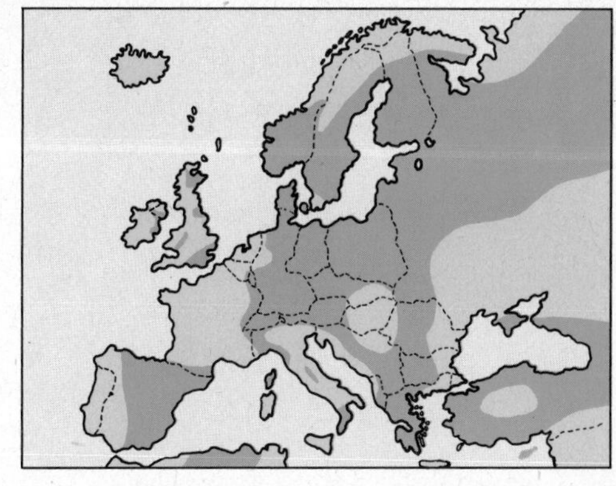

The Crossbill inhabits coniferous forests of spruce, firs and some pines. It feeds mainly on the seeds of spruce cones and other conifers, though it will turn to other seeds such as beech, hornbeam and alder when food is scarce. The bill is specially developed to cope with extracting the seeds from cones. The species is gregarious and is usually seen in small parties. It is acrobatic when feeding and will clamber about in parrot-fashion or hang upside-down like a tit. It makes regular visits to pools to drink water but, apart from when drinking, it is rarely seen on the ground, preferring the tops of trees.
The nest is usually in a tall conifer, 2–25 m from the ground. It is a base of twigs with a cup of grass, moss, lichen and wool lined with hair or feathers. Incubation is by the ♀ alone, fed by the ♂, who also brings food for the young at first. Later both parents bring food.
The Crossbill is not migratory but is erratically dispersive. Sometimes large movements take place, with birds wandering considerable distances. They sometimes breed in the new areas, and as a result the breeding distribution is very sporadic. Most movements take place from late summer onwards and depend upon availability of food and level of population.

The **Two-barred Crossbill** (*Loxia leucoptera*) occurs in north-east Europe and inhabits mainly light larch woods, cedars or mixed conifers. Its habits are much like those of the Crossbill, though it is a rare vagrant in western Europe.

Length:	17 cm
Wing length:	9.5–10 cm
Weight:	40–45 g
Voice:	Explosive 'chup chup'. Song: Greenfinch-like 'ti-chee ti-chee', with creaking and warbling notes
Breeding period:	Very variable. Usually in March/April, sometimes in January/February. 1 brood per year
Size of clutch:	3–4 (2–5) eggs
Colour of eggs:	Light beige ground with a few red-brown patches and twirls, esp. at blunt pole
Size of eggs:	22×16 mm
Incubation:	12–16 days, beginning variable
Fledging period:	Nidicolous; leaving nest at 10–20 days, dependent on parents for further 3–4 weeks

♀ (Li)

t) Ad. ♂ (Li)

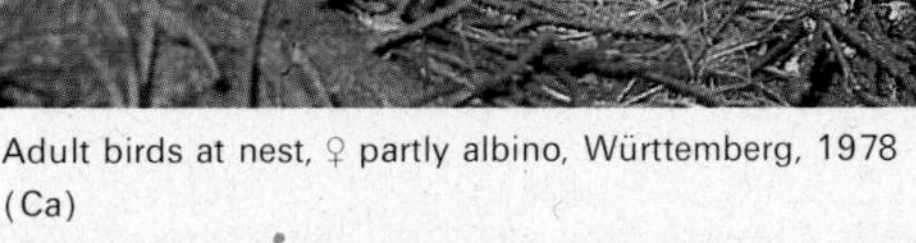

Adult birds at nest, ♀ partly albino, Württemberg, 1978 (Ca)

Baden-Württemberg, 1978 (Schu)

Hawfinch (Coccothraustes coccothraustes)

The Hawfinch inhabits mixed or deciduous woodland with well-grown, mature trees as well as old gardens, orchards, parkland with scattered trees and occasionally in spruce or cedar forest. It is a characteristic bird of the European oak and hornbeam forest. It feeds in the hard kernels and seeds of trees such as cherry, plum, hornbeam, maple and sycamore. It also takes leaf buds and some insects. Much of the time it is found in the tree-tops, where it obtains seeds, but it also feeds on the ground, where it hops in an upright manner. It is a rather quiet and unobtrusive bird. Outside the breeding season it is gregarious and often occurs in flocks. It rarely mixes with other finches. The nest-site is usually on a high horizontal branch of a tall tree, and the nest is a rather untidy structure of twigs, roots and lichen, lined with fine plant materials and hair. Incubation is by the ♀ alone, fed by the ♂. Both parents feed the young on insects and pulped seeds.

The species is largely resident, though some dispersal takes place, determined by availability of food. The more eastern population migrates to winter in the southern part of the range. The movements of this species are irregular and somewhat irruptive.

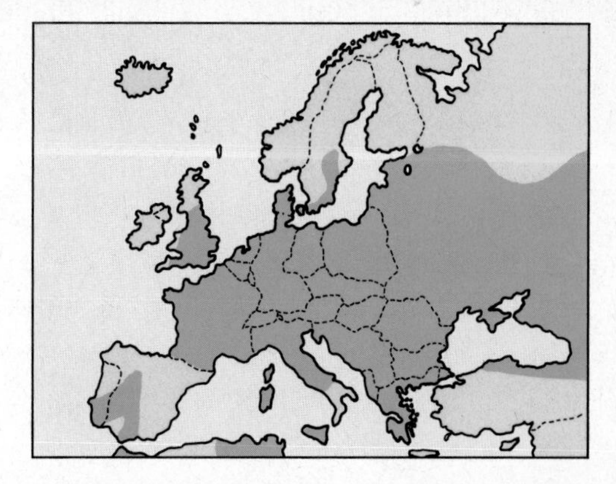

Length:	18 cm
Wing length:	*ca* 10.3 cm
Weight:	*ca* 55 g
Voice:	Sharp 'prik' or 'tzeeip'. Song: like Grosbeak, rather infrequent
Breeding period:	Mid-April to mid-June: 1–2 broods per year. Replacement clutch possible
Size of clutch:	5 (2–7) eggs
Colour of eggs:	Light bluish grey or brownish ground with sparse blackish brown spots, often concentrated at the pole
Size of eggs:	24.5 × 17.5 mm
Incubation:	11–13 days, beg. from complete clutch
Fledging period:	Nidicolous; leaving nest at 10–14 days

ith 9-day-old young, Bavaria, 2.6.1979 (Pf)

Fledged young bird, Württemberg (Schu)

Amperaue, Bavaria, 1968 (Li)

) ♂ winter (Li)

Bullfinch (Pyrrhula pyrrhula)

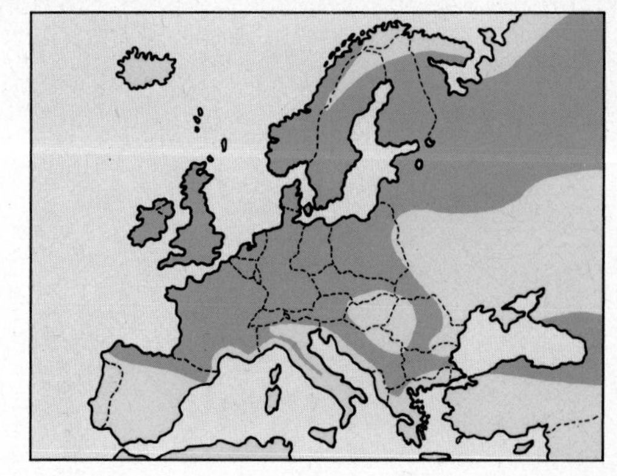

In western Europe the Bullfinch inhabits broad-leaved woodland, parks, gardens with trees or hedges, churchyards and plantations, usually where there is extensive undergrowth of bushes or young conifers. In the rest of the range it is found mainly in coniferous or mixed woodland, often in company with the Crossbill or the Black Woodpecker.
It feeds almost entirely on vegetable matter such as buds, shoots and seeds. Some insects are taken, largely to feed the young. The species is often rather destructive, damaging far more buds than it can eat, and as a result can be a serious pest in fruit-growing districts. It feeds both in the tree-tops and in lower cover or on the ground. It is not gregarious and is usually seen in pairs or family groups, though flocks sometimes occur in early spring.
The nest is a loose structure of twigs, moss and lichen with an inner cup of roots and hair. It is built by the ♀ and is usually placed close to the trunk of a young conifer. Incubation is by the ♀ alone, fed by the ♂. At first the ♂ brings food which the ♀ feeds to the young. Later both parents bring food.
The Bullfinch is mainly resident and sedentary in western Europe. The most northerly and eastern populations tend to disperse southwards in rather irruptive movements rather than in a set migration pattern.
Migration: Main dispersal/irruptions take place in October–November, returning in March–April.

Length:	15 cm
Wing length:	8.5 cm
Weight:	*ca* 30 g
Voice:	Mournful 'dyoo' or 'deu-deu'. Song: a soft warble
Breeding period:	End of April to August. 1–2 (–3) broods per year
Size of clutch:	5–6 (4–8) eggs
Colour of eggs:	Light blue to blue-green, with black or red-brown patches and whorls, concentrated at blunt pole
Size of eggs:	20 × 15 mm
Incubation:	12–14 days, beg. from complete clutch
Fledging period:	Nidicolous; leaving nest at 12–16 days

: nest, Bavaria, 1970 (Li)

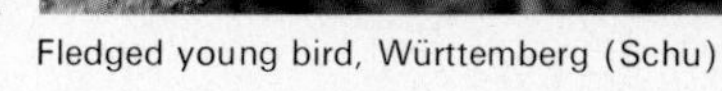

Fledged young bird, Württemberg (Schu)

Upper Bavaria, 22.7.1973 (Pf)

) ♂ at nest, Odenwald, West Germany, 18.5.1963 (Pf)

Snow Finch (Montifringilla nivalis)

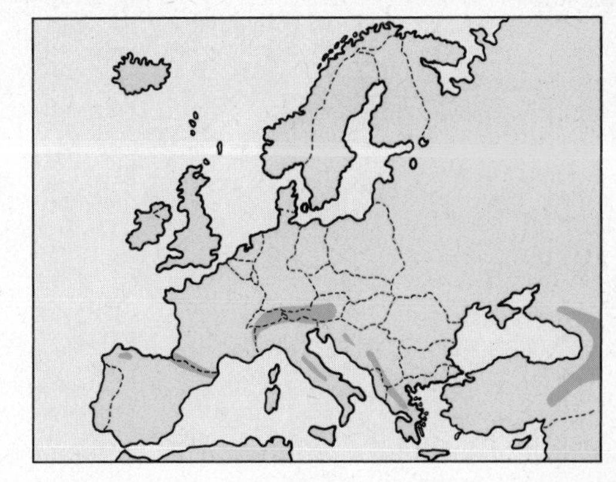

The Snowfinch is found in the mountains of central and southern Europe at a height of 1,300–3,500 m. It occurs on bare, rocky or boulder-strewn slopes often near the snow-line or at the side of glaciers. It also frequents buildings such as alpine huts and in winter descends to high villages and settlements, where it can find food provided by man.
It feeds on small insects, particularly beetles, as well as spiders and a variety of small seeds of alpine plants and the fallen seeds of conifers. It feeds on the ground, searching out insects from among stones and rocks. Outside the breeding season it usually occurs in flocks or small parties and is often very tame and trusting.
The nest-site is in a hole or crevice in rock or among stones, or in a mammal burrow. Holes in walls and buildings are also used where present. The nest is a cup of grass, moss and feathers, lined with feathers and hair. Both sexes incubate and feed the young. The Snow Finch is mainly sedentary, moving to a lower altitude in winter. Occasionally it moves to lowland areas outside the breeding range, but this is rare.

Length:	18 cm
Wing length:	11.5–12 cm
Weight:	*ca* 40 g
Voice:	Hoarse 'tsweehk' or short 'pitch'. Song: repeated 'sittitche-sittitche'
Breeding period:	Mid-May to July. 1–2 broods per year
Size of clutch:	4–5 (3–7) eggs
Colour of eggs:	Pure white, with a dull sheen
Size of eggs:	24×17 mm
Incubation:	13–14 days, beg. before clutch complete
Fledging period:	Nidicolous; leaving nest at *ca* 21 days

feeding young, Austria (Aich)

Austria (Aich)

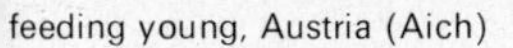

eft) Ad. ♀, Austria (Aich)

Rock Sparrow (Petronia petronia)

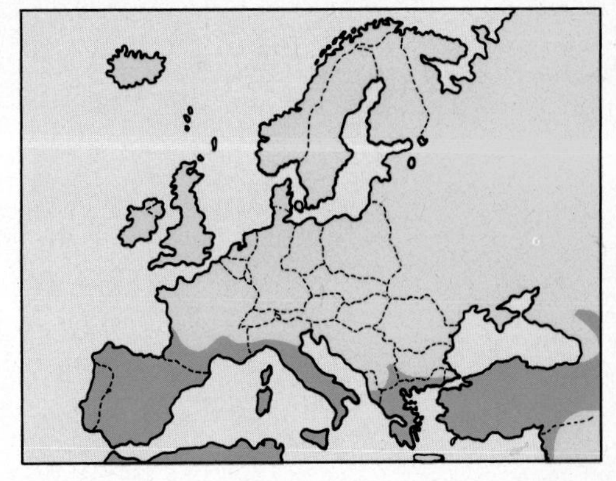

The Rock Sparrow inhabits warm rocky areas in mountains or hilly country, cultivated land (often at high altitude), ruins, walls and large buildings, and towns or villages in some areas. It previously had a larger range in Europe but has retreated southwards, probably due to climatic change.

It feeds on a variety of seeds, grain, buds, shoots and insects. Most of the food is taken on the ground where the species runs quickly among stones and boulders. It also perches in trees and will take some food there. It is gregarious outside the breeding season and often mixes with House Sparrows.

The nest-site is in a hole or crevice among rocks or stone as well as in hollow trees, rodent burrows and holes in walls or buildings. The nest is a somewhat domed structure of grasses, plant stems and roots lined with hair and feathers. The ♀ takes the main or sole role in incubation but both parents feed the young.

The Rock Sparrow is largely sedentary, remaining close to the nesting areas throughout the year. In the east of the range some dispersal takes place in winter, and the species has occurred as a vagrant in north European countries, including Britain.

Length:	14 cm
Wing length:	9.7 cm
Weight:	*ca* 35 g
Voice:	Characteristic 'tut'; also 'chwee' or 'pey-i'. Song: Sparrow-like
Breeding period:	Mid-April to June. 1–2 broods per year
Size of clutch:	5–6 (4–7) eggs
Colour of eggs:	Whitish to greenish-white ground, sprinkled with red-brown or dark brown spots, esp. at blunt pole
Size of eggs:	22×15.5 mm
Incubation:	12–14 days, beg. from complete clutch
Fledging period:	Nidicolous; leaving nest at *ca* 21 days

d. Turkey, 9.6.1977 (Li)

Nestlings, turkey, 17.6.1977 (Li)

Turkey, 9.6.1977 (Li)

ꞁft) Ad. with prey, Greece, 12.6.1975 (Pf)

House Sparrow (Passer domesticus)

The House Sparrow is closely associated with human habitation and is usually found in the vicinity of towns and villages or in cultivated land, farms and settlements. It occurs in the centre of large cities and is one of the most familiar European birds.
The diet is very varied, and includes vegetable and animal food-scraps discarded by man as well as seeds, grain, shoots, buds, insects and spiders. It feeds mainly on the ground but will take buds and seeds in trees and bushes. It roosts socially and is gregarious for much of the year. Large flocks occur on agricultural land, sometimes in company with finches and buntings. These large congregations of sparrows constitute an agricultural pest in some areas. The nest-sites are very varied, ranging from holes in walls or trees to multiple nests in trees and in the nest of large birds of prey or storks. It will often drive other birds away from their nests and take them over for their own use. This is frequent with the House Martin, the Swallow and the Swift. The nest is a large untidy domed structure of straw, plant stems, discarded rubbish and similar materials, lined with feathers, wool or hair. Both sexes take part in nest-building but it is mainly done by the ♂. The ♀ takes the major role in incubation and both parents feed the young.
The species is largely resident and sedentary.
In Italy it is replaced by a race called the **Italian Sparrow** (*P. d. italiae*) which has similarities to the Spanish Sparrow.

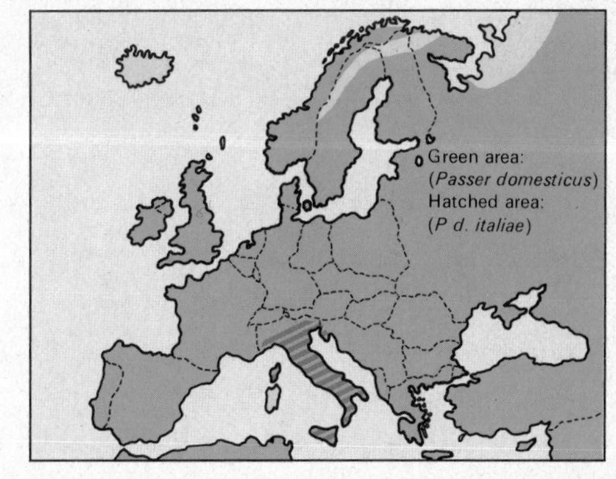

Length:	15 cm
Wing length:	8 cm
Weight:	*ca* 30 g
Voice:	'Cheep' and 'chissis'; also many twittering and chirping notes
Breeding period:	Usually from mid-April to end of August. 2–3 broods per year
Size of clutch:	5–6 (3–8) eggs
Colour of eggs:	Very variable. Dense grey or brown patches on a white to greenish ground
Size of eggs:	22.5×16 mm
Incubation:	12–14 days, beg. from complete clutch
Fledging period:	Nidicolous; leaving nest at *ca* 14 days

use Sparrow ♀, Bavaria, June 1978 (Pf)

Italian Sparrow ♂, Upper Bolzano, 1977 (Ta)

House Sparrow's nest, Bavaria, 8.8.1978 (Pf)

ft) House Sparrow ♂, Bavaria, June 1978 (Pf)

Spanish Sparrow

(Passer [domesticus] hispaniolensis)

The Spanish Sparrow is found in areas of low scrub and thickets along river banks and in dried-up river beds. It also inhabits olive groves and sparse wooded areas. In some regions the habitat is similar to that of the House Sparrow, including human settlements, but it generally avoids cultivated land. It sometimes hybridises with House Sparrows, usually in areas near human habitation.

The diet is much the same as the House Sparrow but less human refuse is taken due to the difference in habitat.

The nest-site is usually in trees or bushes, rarely in a wall or building. The nests of storks and other large birds are also used when these are in natural sites. It usually nests colonially with many nests in the same tree. The nest is larger than that of the House Sparrow, being a bulky domed structure with a side-entrance. It is built of plant stems, straw and grass, lined with feathers, hair or wool. The breeding biology is very similar to that of the House Sparrow.

Spanish Sparrows are mainly migratory, wintering in Africa north of the Sahara and down the Nile Valley to the Sudan. It also winters in the Middle East and Arabia. Hybrid populations are resident and sedentary. The species has occurred as a vagrant in areas outside the normal range, including Britain.

Migration: Autumn movements mainly late September to October, returning in March–April.

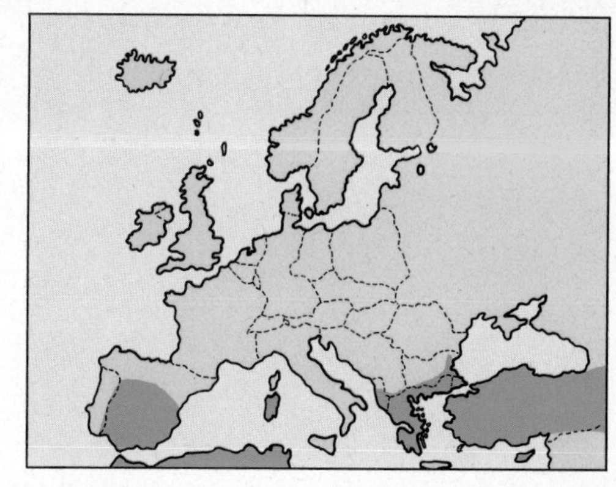

Length:	15 cm
Wing length:	7–9 cm
Weight:	*ca* 30 g
Voice:	'Chup', like House Sparrow, but rather deeper
Breeding period:	April–June. 2 (–3) broods per year
Size of clutch:	5–6 (4–8) eggs
Colour of eggs:	Like eggs of House Sparrow, but not so thickly marked
Size of eggs:	22 × 15.5 mm
Incubation:	12–13 days, beg. from complete clutch
Fledging period:	Nidicolous; leaving nest at *ca* 12–15 days

. ♂, Greece, 4.7.1977 (Sy)

Juv., Greece, 5.6.1978 (Li)

Greece, 16.5.1972 (Li)

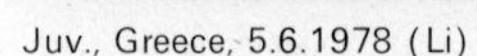

ɩ) ♀ at nest, Greece, 5.6.1978 (Li)

Tree Sparrow (Passer montanus)

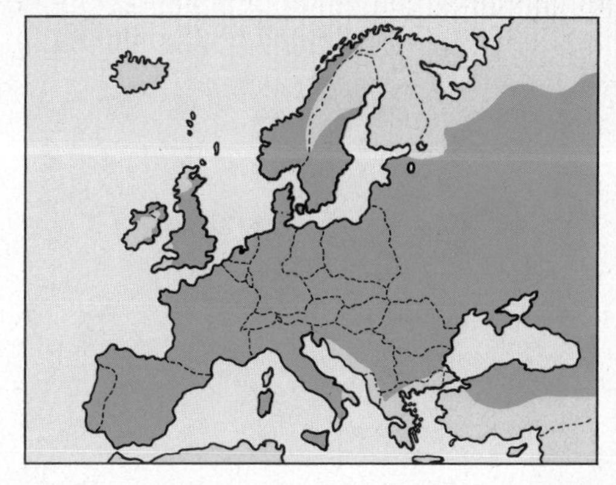

In Europe the Tree Sparrow is found in open areas with large old trees, hedgerows, pollarded willows along river-banks, quarries, banks, ruins and old buildings as well as orchards and parkland. It is not closely attached to human habitation, though it does occur in suburban areas and in Asia, occupying the same niche as the House Sparrow.
It feeds on seeds and grain as well as on insects and their larvae. Most of the food is obtained on or near the ground. The species is gregarious outside the breeding season and often occurs in flocks or mixed with finches and House Sparrows. It frequents stubble-fields, stackyards and farmland in winter.
The nest is usually in a hole in a tree, cliff or quarry, sometimes in company with Sand Martins on riverbanks. It also uses mammal-holes, artificial sites such as pipes or nest-boxes and is found among creepers. The nest is an untidy structure of plant stems and twigs lined with feathers and down. Both sexes take part in nest building, incubation and care of the young.
In areas where it comes into contact with House Sparrows in the breeding season occasional hybrids are known.
The species is largely sedentary, though it wanders in winter. Some northern birds migrate to the southern part of the range in winter. Movements are sometimes irruptive and may result in birds colonising a new area, often for only a few years.

Length:	14 cm
Wing length:	6.5–7.5 cm
Weight:	*ca* 23 g
Voice:	Calls shorter and quicker than House Sparrow: 'chickchick-chick'. Flight-call: 'tek-tek'
Breeding period:	April–July. 2–3 broods per year
Size of clutch:	5–6 (2–10) eggs
Colour of eggs:	Light ground, densely covered with dark brown or grey patches
Size of eggs:	19.5 × 14 mm
Incubation:	12–14 days, beg. from complete clutch
Fledging period:	Nidicolous; leaving nest at *ca* 12–14 days

aria, 8.11.1975 (Li)

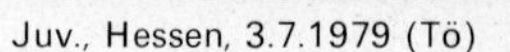

Juv., Hessen, 3.7.1979 (Tö)

Bavaria, 3.6.1967 (Li)

*) Pair at nest-hole, Bavaria, 14.5.1979 (Li)

Rose-coloured Starling

(Sturnus roseus)

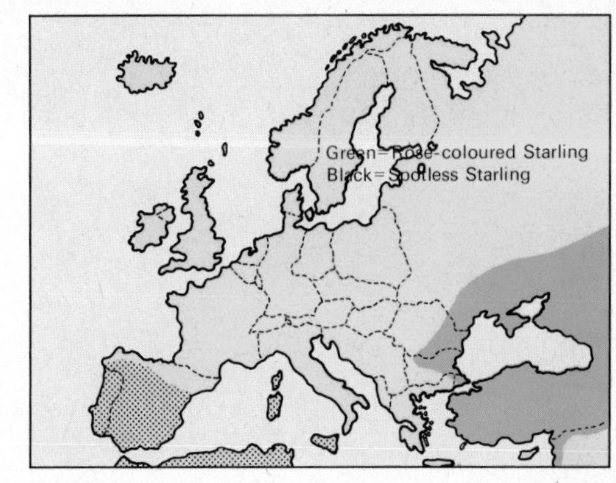

The Rose-coloured Starling inhabits dry grassy steppes and rocky hillsides, open cultivated land and old buildings.
It feeds on grasshoppers, locusts, large beetles and other insects as well as fruit and berries. Much of the food is obtained on the ground but it also feeds in trees and hawks for insects on the wing. It is a sociable bird. It mixes freely with Starlings and is often seen in association with cattle and other livestock. The nest is usually in a hole in the ground or among stones and rocks. The nest is an untidy collection of plant stems, grass and leaves with a lining of hair and feathers. The ♀ incubates alone and the young are fed by both parents.
The species is migratory, wintering mainly in northern India. It is prone to mass irruptions, usually following migratory locusts. The nesting range varies considerably and may extend into western European countries such as Hungary and Italy at times.

Spotless Starling (*Sturnus unicolor*) The Spotless Starling has a distribution complementary to that of the Starling. It closely resembles that species in choice of habitat, diet, habits and breeding biology, though the ♀ usually incubates alone. The species is partially migratory, wintering in North Africa, though some remain in the breeding areas.

Length:	21.5 cm
Wing length:	12.4–13.5 cm
Weight:	*ca* 76 g
Voice:	Starling-like 'chuurr'. Song: Starling-like with high pitched musical and chattering noises
Breeding period:	Beginning of May to June. 1 brood per year
Size of clutch:	5–6 (3–9) eggs
Colour of eggs:	Light bluish to bluish-white, smooth and shiny
Size of eggs:	28.5×21 mm
Incubation:	11–14 days, beg. before clutch complete
Fledging period:	Nidicolous; leaving nest at *ca* 14–19 days

Spotless Starling, Spain (Haas)

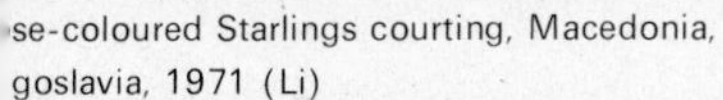

se-coloured Starlings courting, Macedonia,
goslavia, 1971 (Li)
ft) Ad. ♂, Macedonia, Yugoslavia, 1971 (Li)

Starling (Sturnus vulgaris)

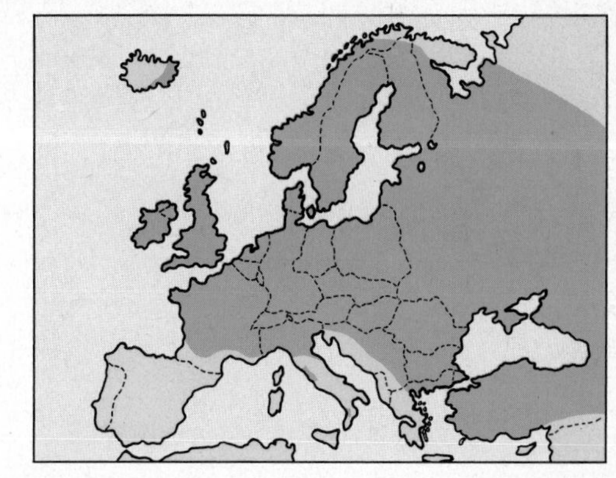

The Starling is found in a wide variety of habitat where suitable nest-sites occur. It inhabits farmland, open country with scattered trees, woodland, quarries and sea-cliffs, ruins, towns and cities. It feeds on a variety of insects, earthworms, spiders, woodlice, snails and slugs as well as grain, seeds, berries, fruit and other vegetable matter. Much of the food is obtained on the ground, but it will also hawk for flying insects. It is sociable at all times, roosting in large numbers. Often flocks perform mass aerial evolutions before diving down to the roost.

The nest-site is in a hole in a tree, wall or cliff, sometimes close to the ground. It also nests in sheds and buildings and takes to nest-boxes readily. The nest is an untidy mass of stems, leaves and other plant material lined with feathers, wool or moss. The ♂ builds the nest before pairing, and then the ♀ finishes the lining. Both sexes incubate and feed the young, which are dependent on the parents for some time after fledging.

In southern and western Europe the Starling is largely resident. The eastern and northern populations are entirely migratory, wintering in temperate Europe, North Africa and the Middle East. Very large flocks occur on migration, and the species can travel vast distances.

Migration: Main autumn movement take place in late September to mid-November, with return passage in mid-February to late April.

Length:	22 cm
Wing length:	12.2–13.2 cm
Weight:	*ca* 75 g
Voice:	Usual note 'tcheerr'. Song: rambling mixture of whistling, warbling and chattering, often strangely imitative
Breeding period:	Mid-April to July. 1–2 broods per year
Size of clutch:	5–6 (4–9) eggs
Colour of eggs:	Uniform blue-green to light blue
Size of eggs:	30×22 mm
Incubation:	12–15 days, beg. from complete clutch
Fledging period:	Nidicolous; leaving nest at 20–22 days

nter plumage, Bavaria, 1978 (Li)

Nestlings, Bavaria, May 1975 (Li)

Bavaria, 5.4.1976 (Li)

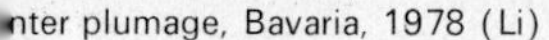

t) Breeding plumage, Bavaria, April 1979 (Li)

Golden Oriole (Oriolus oriolus)

The Golden Oriole inhabits open woodland, parkland with trees, orchards, plantations and trees bordering roads. It is usually found in warm, sunny locations and favours tall, mature broad-leaved trees. It is not found in coniferous forest.
It feeds on large insects and larvae of tree-dwelling species, as well as grasshoppers, bees, spiders and small snails. In autumn it takes a great amount of fruit and berries, including cherries, grapes, figs and mulberries. Much of the food is obtained in the tree-tops but it will also feed in lower vegetation and on the ground. It is not sociable, and is usually seen singly or in pairs. There is a courtship flight which involves dashing at full speed through the trees, the ♂ following the ♀ closely, and conforming to her every movement. The nest is a remarkable structure; a cup of grasses and stems with strips of bark woven around and suspended from a forked branch. The nest is lined with wool or fine grass and is built by the ♀. Both sexes incubate but the ♀ takes the larger share, and both parents feed the young.
The Golden Oriole is entirely migratory, wintering in tropical Africa, mainly south of the Equator. It occurs on passage in small numbers in Britain and has bred on occasions.
Migration: Autumn movements in late August to September, returning in late April to early June.

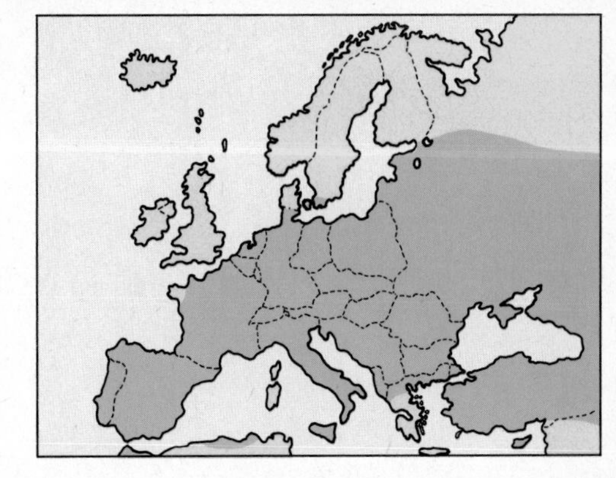

Length:	24 cm
Wing length:	14.5–16 cm
Weight:	*ca* 72 g
Voice:	Melodious 'weela-weeoo'. Warning call: a scratching 'krraah'
Breeding period:	End of May to July. 1 brood per year. Replacement clutch possible
Size of clutch:	3–4 (6) eggs
Colour of eggs:	Whitish to light pink with a few purplish or dark brown spots
Size of eggs:	30.5×21 mm
Incubation:	14–15 days, beg. from complete clutch
Fledging period:	Nidicolous; leaving nest at 14–15 days

Bavaria, June 1965 (Li)

Young bird, Amperaue, Bavaria, 25.7.1977 (Köhn)

Upper Bavaria, 29.5.1976 (Pf)

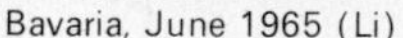

eft) ♂, Amperaue, Bavaria, 28.6.1972 (Li)

Siberian Jay (Perisoreus infaustus)

The Siberian Jay breeds in the dark spruce taiga forests of the north. It also occurs in some mixed or birch woods and, in winter, resorts to outskirts of settlements and villages.

It feeds on seeds of pines and cedars, berries, small mammals, young birds and eggs, insects and all manner of edible substances obtained around human habitation. It is rather shy and retiring in the breeding season but in winter becomes very tame and trusting. Much of the food is obtained on the ground but it will climb out to the tips of branches to take seeds. The nest is usually sited in a conifer, close to the trunk and between 2 m and 12 m from the ground. It is a thick cup of twigs and bark lined with lichen and feathers. The ♀ incubates alone, fed by the ♂, who also brings all the food for the young at first. Later both parents bring food.

The species is mainly resident and sedentary, though local movements occur in winter and occasionally it appears outside the normal breeding range. The range in Scandinavia has retreated northwards in the present century, probably as a result of climatic changes.

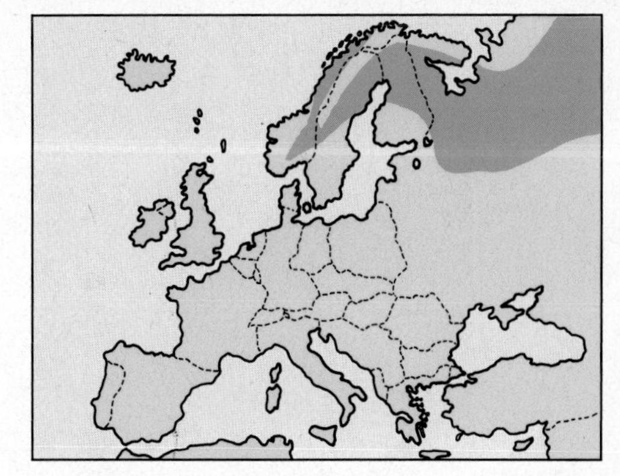

Length:	30 cm
Wing length:	14 cm
Weight:	*ca* 155 g
Voice:	Harsh Jay-like sounds, e.g. 'chair' and cheerful 'cook-cook'
Breeding period:	Mid-April to mid-May. 1 brood per year. Replacement clutch possible
Size of clutch:	3–4 (5–6) eggs
Colour of eggs:	Bluish ground with brown spots, sometimes concentrated at blunt pole
Size of eggs:	30×22 mm
Incubation:	18–20 days, beg. from 1st egg
Fledging period:	Nidicolous; leaving nest at 21–23 days, remains in family group through winter

d., Jämtland, Sweden, 29.6.1963 (Chr)

Fledged young bird, north-east Finland, 18.6.1978 (Gr)

Finland (Hau)

ft) Ad. at nest, Finland (Hau)

Jay (Garrulus glandarius)

The Jay inhabits deciduous and mixed woodlands, open wooded areas in parks, plantations, orchards and sometimes hedgerows. It also occurs in light coniferous woodland in montane and northern areas.
It feeds on large insects, especially caterpillars and beetles, seeds and fruits of all types including acorns, beechmast, peas, grain and pine seeds as well as young birds and eggs of small- to medium-sized woodland birds, mice, frogs and earthworms. It is mainly aboreal but also feeds on the ground. Outside the breeding season it often occurs in pairs or small groups and larger flocks are sometimes recorded.
It nests in trees from about 2 m above the ground. The nest is usually sited in a fork or against the trunk and is well hidden. It is a cup of twigs and plant stems, often with some earth as a binding material. It is lined with fine plant matter and hair to form a neat inner cup. Incubation is shared by both parents, and both bring food for the young, though at first the ♀ broods and the ♂ brings all the food.
The Jay is mainly sedentary, though birds from the northern coniferous zone make sporadic irruptive movements depending upon the availability of food. The species has extended its range to the north in the present century.

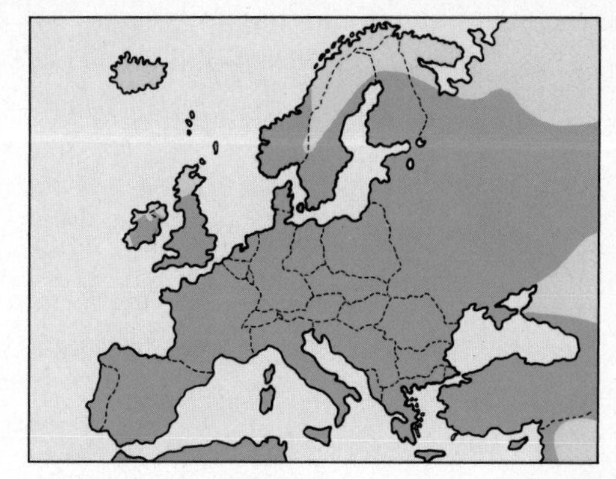

Length:	34 cm
Wing length:	18.5 cm
Weight:	*ca* 170 g
Voice:	Harsh 'kaark-kaark'; also prolonged mewing note
Breeding period:	End of April to mid-June. 1 brood per year. Replacement clutch possible
Size of clutch:	5–6 (3–10) eggs
Colour of eggs:	Light olive-green to sandy ground, with even light brown spots
Size of eggs:	31.5 × 23 mm
Incubation:	16–17 days, often beg. from 1st egg
Fledging period:	Nidicolous; leaving nest at 19–20 days

per Bavaria, 1.11.1977 (Pf)

Young bird sunbathing, Bavaria, 1969 (Li)

Bavaria, 21.4.1972 (Li)

ft) Bavaria, 5.5.1973 (Li)

Azure-winged Magpie

(Cyanopica cyanus)

The Azure-winged Magpie has an extraordinary distribution, being found in Iberia and in China, Japan and east Asia.
In Europe it is found in light oak-woods, stone pines and mixed woodland as well as orchards, olive groves and dry scrub with scattered trees.
It feeds on large insects, berries, fruits and small animals, with most of the food being taken on the ground or in low bushes. The birds frequently form roaming parties, moving noisily through eucalyptus groves and other wooded areas. In the breeding season it is more secretive. In general habits it resembles the Magpie, and is quick and alert in its movements.
The nest is usually in a stone pine tree, placed in a fork about 3–5 m from the ground. It is a bulky cup of twigs, roots and other plant matter, sometimes with some mud as binding, and lined with plant fibres, hair and wool. Incubation is by the ♀ alone. Both parents feed the young on regurgitated food.
The species is entirely sedentary, though it roams freely within the breeding areas. Birds often remain in family groups through the winter.

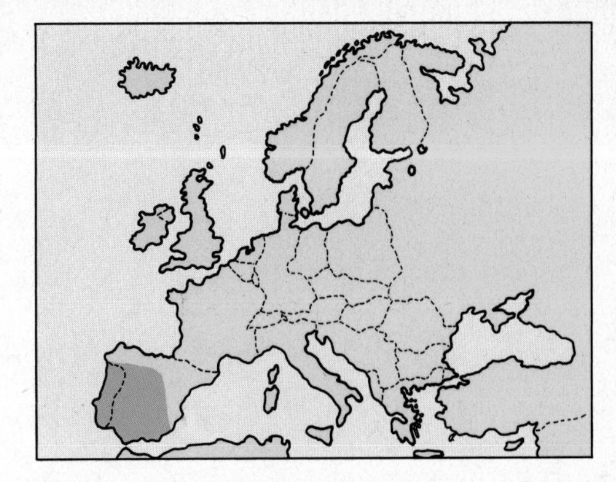

Length:	34 cm
Wing length:	14 cm
Weight:	75 g
Voice:	Often repeated, very varied calls of 'zhree' or 'kwear', 'quit-quit-quit', etc.; very vocal
Breeding period:	End of April to May/June. 1 brood per year. Replacement clutch possible
Size of clutch:	5–6 (–9) eggs
Colour of eggs:	Cream to light olive ground with sparse brown spots, concentrated at blunt pole
Size of eggs:	28×20 mm
Incubation:	Unknown
Fledging period:	Unknown

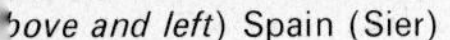

bove and left) Spain (Sier)

Spain (Haas)

Spain, 25.4.1979 (Br)

Magpie (Pica pica)

The Magpie is found in a variety of woodland habitats including broad-leaved, mixed and coniferous forest, orchards, hedgerows, scattered trees in open country, areas of low dense scrub and bushes, and sometimes near habitation in virtually treeless country.
The varied diet includes insects, mainly large species of beetles and caterpillars, small mammals, eggs and young birds, roadside casualties and other carrion, seeds, grain, nuts, fruit and berries. Much of the food is obtained on the ground and surplus food is sometimes hoarded.
The species often occurs in small groups and is generally sociable, with large flocks congregating in winter.
The nest is a bulky structure of sticks and mud with a neater inner cup of roots, plant fibres and hair. It is covered with an open dome of twigs leaving an opening at the side and is sited in the top of a bush or tall tree, rarely on man-made structures such as pylons.
The Magpie is mainly resident and sedentary.
Both sexes build the nest with materials brought by the ♂. Incubation is by the ♀ only and the young are fed by both parents.

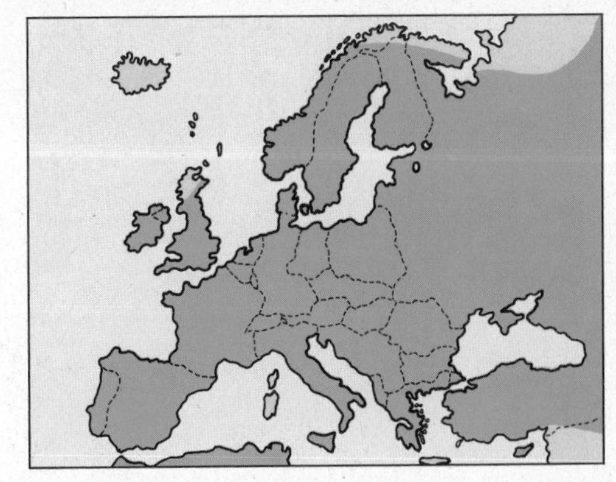

Length:	45 cm
Wing length:	17–20 cm
Weight:	*ca* 210 g
Voice:	Quick 'shack-shack-shack . . .' or 'shirrrack'
Breeding period:	End of March to June. 1 brood per year. Replacement clutch possible
Size of clutch:	6–8 (4–10) eggs
Colour of eggs:	Grey-green to bluish ground, with olive-brown spots
Size of eggs:	33.5×24 mm
Incubation:	17–18 days, usually beg. with first egg
Fledging period:	Nidicolous; leaving nest at 22–27 days

enmark (Ge)

Ad. at nest, Bavaria, 6.4.1978 (Li)

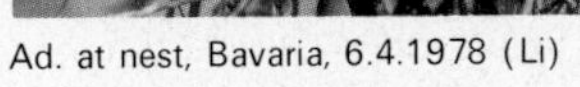

Upper Bavaria, 26.4.1973 (Pf)

eft) Bavaria, 1975 (Li)

Nutcracker (Nucifraga caryocatactes)

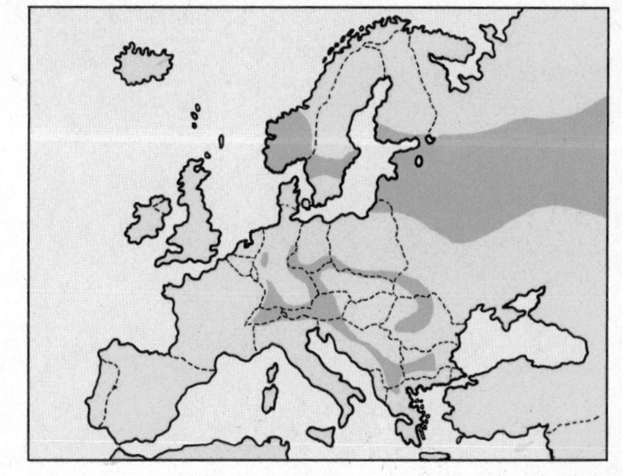

The Nutcracker inhabits coniferous forests of all types as well as mixed deciduous woodland with some conifers. In the northern part of the range it is found in the taiga forest and in central Europe mostly in mountain pines and larches. Two races occur, the Thin-billed form (*N. C. macrorhynchos*), which breeds in Siberia and European Russia, and the thick-billed (nominate) form, which inhabits the mountains of southern and central Europe.
The diet includes conifer seeds, nuts and seeds of beech, hazel, etc., as well as insects, earthworms, eggs and the young of small birds, berries and carrion. It feeds on the ground, sometimes digging in the snow to find food, and in the tops of trees to find cones. Surplus food is sometimes hoarded. Outside the breeding season it often occurs in small parties.
The nest is usually in a conifer, close to the trunk and fairly high above the ground. It is a large cup of twigs and bark lined with lichen and grass to form a thick layer. Incubation is by the ♀ alone and the young are fed by both parents, remaining dependent on them for 2–3 months after fledging.
The central European race is mainly resident but the thin-billed Siberian form often erupts in mass movements. These movements take the birds as far west as the Pyrenees and sometimes result in that form breeding well outside its normal range.
Migration: Irruptions usually start in late summer with progressive movement through the winter pressing further west.

Length:	32 cm
Wing length:	18.5 cm
Weight:	*ca* 200 g
Voice:	Jarring 'krohkrohkroh' or Jay-like 'kraak'
Breeding period:	Mid-March to May. 1 brood per year. Replacement clutch possible
Size of clutch:	3–4 (2–5) eggs
Colour of eggs:	Very variable. Generally brownish spots on whitish or bluish ground
Size of eggs:	34×24 mm
Incubation:	17–19 days, beg. with 1st or 2nd egg
Fledging period:	Nidicolous; leaving nest at 21–28 days

d. of race *N. C. macrorhynchos*

Falsterbo, Sweden, 18.9.1975 (Chr)

Ad. at nest, Sweden (Sw)

ft) Ad. extracting pine-kernels

Alpine Chough (Pyrrhocorax graculus)

The Alpine Chough inhabits the alpine zone of southern and central Europe. It is found right up to the snow-line on cliffs and rocky walls and in grassy alpine meadows. The diet is mainly large insects, earthworms, molluscs and small animals. It is also a scavenger. It feeds on the ground. It is expert at soaring and gliding in the air-currents above the mountains. It usually occurs in flocks. It also nests colonially.
The nests are sited in holes or crevices in rock-faces, often deeply hidden. The nest is a bulky cup of twigs and roots with a lining of finer plant material. The ♀ incubates alone, fed by the ♂.
The species is sedentary; but moves to a lower altitude in winter, though it rarely descends to lowland areas.

Chough (Pyrrhocorax pyrrhocorax) The Chough inhabits sea-cliffs, quarries, uplands and mountainous areas, though generally at a lower altitude than the Alpine Chough. The diet is similar but includes crustaceans and other shore-line animals. It nests in holes, crevices or caves and sometimes in old buildings. The nests and breeding biology are similar to that of the Alpine Chough. It is generally more widespread except in the high mountains of central Europe. It is resident and sedentary.

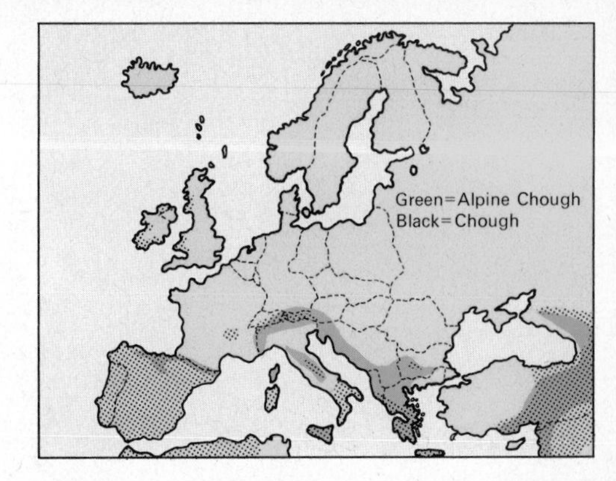

Length:	38 cm — Alpine Chough
Wing length:	27 cm
Weight:	240 g
Voice:	Loud, rippling 'chirrich', and short, loud calls of 'tchiyup'
Breeding period:	April–June. 1 brood per year. Replacement clutch possible
Size of clutch:	4–5 (3–6) eggs
Colour of eggs:	Bright brown spots on whitish or reddish ground
Size of eggs:	39×26.5 mm
Incubation:	18–20 days, beg. from complete clutch
Fledging period:	Nidicolous; leaving nest at *ca* 4 weeks

Length:	38–40 cm — Chough
Wing length:	*ca* 29 cm
Weight:	*ca* 250 g
Voice:	Jackdaw-like 'kyaw', also 'kwuk-uk-uk'
Breeding period:	End of April to June. 1 brood per year. Replacement clutch possible
Size of clutch:	3–4 (2–7) eggs
Colour of eggs:	Cream to light green with small grey and olive-brown spots and patches
Size of eggs:	40×28 mm
Incubation:	17–23 days, beg. from 1st egg
Fledging period:	Nidicolous; leaving nest at *ca* 38–40 days

d. Chough, Zoo photograph (Li)

Imm. Chough, Turkey, 1977 (Wa)

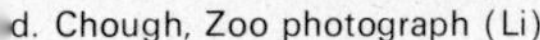

eft) Alpine Chough, Bavaria, 12.2.1975 (Li)

Jackdaw (Corvus monedula)

The Jackdaw is found in open woodland with mature trees which offer suitable nest-sites, areas of scattered trees in farmland or open country, sea- and inland cliffs, quarries, ruins and old buildings, and in towns and cities. The varied diet includes many insect larvae, spiders, earthworms, seed, nuts, grain, all manner of discarded human refuse, carrion, etc. Much of the food is obtained on the ground, often in grassy fields, farmland and refuse dumps. It will also take food from trees and larger bushes. It is a sociable bird, usually seen in groups or large flocks. It often nests colonially in groups of 1–2 pairs to quite large numbers. It will associate with other corvids such as Crows and Rooks as well as Starlings.

The nest-site is in a hole in a tree or hollow tree, holes and cracks in rock faces, ruins or large buildings and sometimes in the chimneys of occupied houses. The nest is a base of twigs which may be substantial at times, with accumulated débris forming a massive pile, blocking chimneys or filling holes in trees. At other times it is reduced to a few twigs. It is lined with wool, hair or other débris. Incubation is by the ♀ alone, with both parents feeding the young.

The Jackdaw is mainly resident and sedentary, though birds from eastern and northern Europe move south and westwards in winter. Outside the breeding season it often forms large flocks which may wander from the breeding colonies when food is scarce. The species has expanded its range northwards into Fenno-Scandia in the present century.

Migration: Autumn movements mainly October–November, returning in February–April.

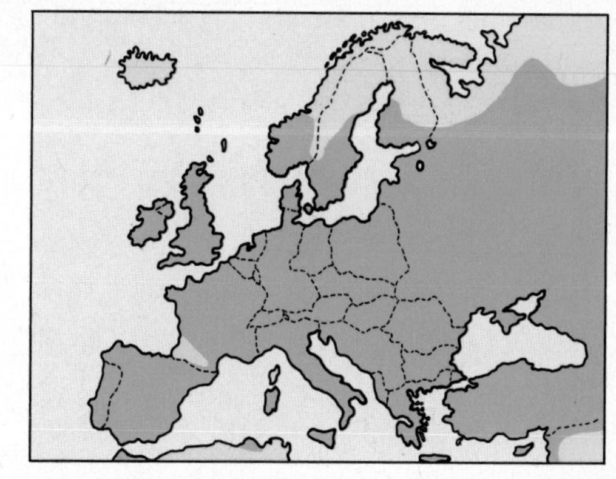

Length:	33 cm
Wing length:	24 cm
Weight:	230 g
Voice:	Resonant 'chack', repeated several times; also 'kya'
Breeding period:	Mid-April to May. 1 brood per year. Replacement clutch possible
Size of clutch:	4–6 (3–7) eggs
Colour of eggs:	Light blue ground with fairly strong markings. Spots denser at blunt pole
Size of eggs:	35×25 mm
Incubation:	17–18 days, beg. varies acc. to individual
Fledging period:	Nidicolous; leaving nest at *ca* 28 days

d. east European form (*C. m. soemmerringii*) (Pf)

Fledged young bird, Greece, 17.6.1976 (Li)

Yugoslavia, 1978 (Li)

ft) Ad., Denmark (Chr)

Rook (Corvus frugilegus)

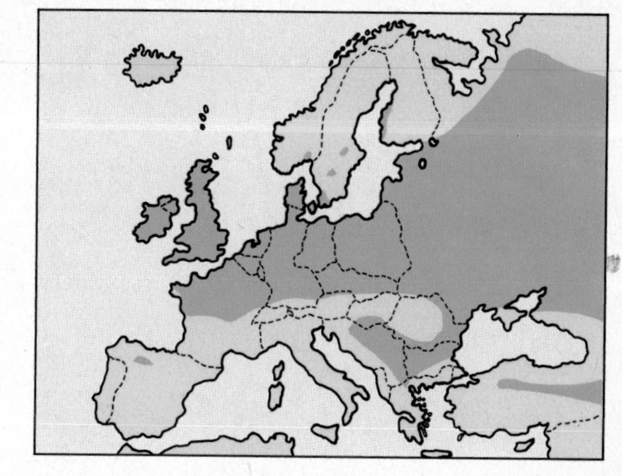

Rooks are found mainly in open country such as farmland with groups of tall trees. It avoids dense woodland though it occurs in copses with tall trees or on the edge of open woodlands. It is often found in association with human settlements, including the suburbs of large towns.
The diet includes grain, seeds, earthworms, insects, larvae, scraps and sometimes carrion. It can be injurious to agriculture but also takes many harmful insect pests. It feeds mainly on the ground, typically in cultivated fields or short grassy areas. It is essentially sociable, nesting colonially and keeping in groups or larger flocks throughout the year.
The nests are placed in tall trees, often elms, usually in a fork in the branches. The nest is a platform of twigs and may be used in successive years, so that an accumulation of nesting material results. Many nests may be built in the same tree and, as nesting takes place before the foliage appears, they are very noticeable. Incubation is by the ♀ alone with the young being fed by both parents. In western Europe the Rook is mainly resident and sedentary though it may disperse in search of food after the breeding season. Birds from central and eastern Europe are largely migratory, wintering in the west of the range and in Mediterranean countries and Asia Minor. It forms foraging flocks in company with Starlings and other corvids.
Migration: Autumn movements mainly October–November, returning in February–March.

Length:	46 cm
Wing length:	32 cm
Weight:	330–670 g
Voice:	Deep 'kaaa' or 'kew', less harsh than Carrion Crow
Breeding period:	End of March to May. 1 brood per year. Replacement clutch possible
Size of clutch:	4–5 (3–9) eggs
Colour of eggs:	Light blue to greenish ground, with brownish spots. Colour often varies within a clutch
Size of eggs:	40 × 28.5 mm
Incubation:	16–20 days, usually broods from 1st egg
Fledging period:	Nidicolous; leaving nest at 30 days

varia, 15.2.1978 (Li)

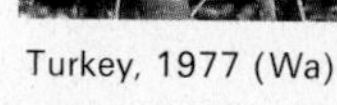

Turkey, 1977 (Wa)

Bavaria, 29.4.1975 (Li)

t) Colony of Rooks, Skåne, Sweden, 6.5.1971 (Chr)

Carrion Crow (Corvus corone corone)

The Carrion Crow forms one species with the Hooded Crow and the two forms freely interbreed where their ranges meet. It is found in a wide range of open country including farmland, heath and moorland and upland areas as well as parks, towns and cities. It also occurs on cliffs and sea-coasts, though the Hooded Crow is more frequent in these habitats when in range. It feeds mainly on the ground and takes a wide assortment of insects, plant matter, scraps and refuse, eggs and young of other birds, carrion of all types including road-casualties and dead livestock. It is not a social bird like the Rook. It nests singly and is usually seen in pairs or family parties, though numbers may associate together when feeding and at roosts.
It nests in trees, usually in the fork of a branch, and on pylons and other man-made structures, ledges, cliffs and, in some places, on the ground in deep heather or other vegetation. The nest is a large structure of sticks, plant stems, kelp fronds at coastal sites, mixed with some earth and moss or heather. The nest is lined with wool or hair and is often decorated with bleached bones. Nests may be used in successive years and become quite substantial. Incubation is by the ♀ alone, with both parents feeding the young.
The Carrion Crow is almost entirely resident and sedentary, though occasionally it wanders away from the breeding range. It is subject to persecution in some areas because of the effect its predation has on stocks of game birds.

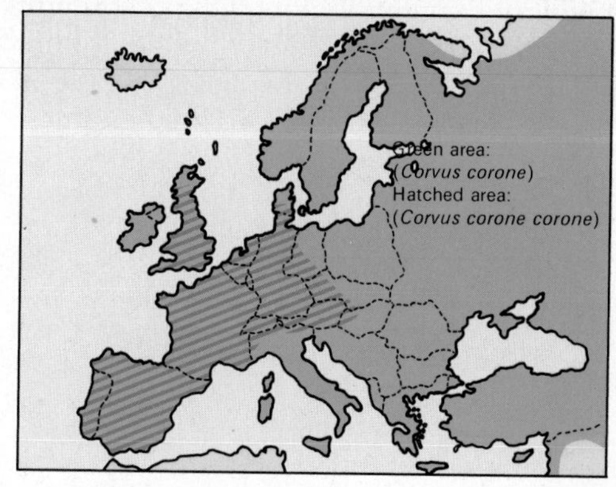

Length:	47 cm
Wing length:	33 cm
Weight:	400–800 g
Voice:	Harsh, croaking 'kree', often repeated; also 'keerk' and 'konk'
Breeding period:	Beginning of April to May. 1 brood per year. Replacement clutch possible
Size of clutch:	5 (4–7) eggs
Colour of eggs:	Light blue to light green ground with brown spots, often denser at blunt pole
Size of eggs:	42.5×30 mm
Incubation:	18–20 days, beg. varies acc. to individual
Fledging period:	Nidicolous; leaving nest at *ca* 4 weeks

varia, April 1974 (Pf)

Nestlings, Bavaria, 29.4.1975 (Li)

Bavaria, 3.4.1971 (Li)

t) Bavaria (Diep)

Hooded Crow (Corvus corone cornix)

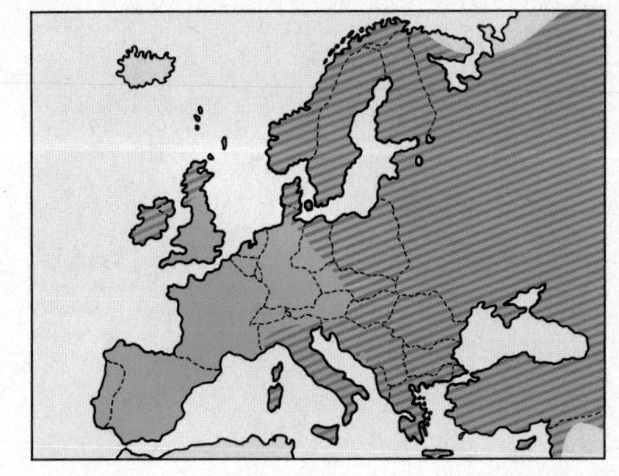

The Hooded Crow inhabits country similar to that of the Carrion Crow, though, because of the difference in range, it is found more frequently on rocky sea-cliffs, open moor and heathland and bare mountainous country. The diet, habits and breeding biology are the same as for the Carrion Crow though the diet varies due to the difference in range. The Hooded Crow is perhaps more sociable than the Carrion Crow, due in part to its migratory tendencies.

Where the ranges of Hooded and Carrion Crows meet there is a zone of hybridisation which extends up to 100 km. The fact that the width of this zone is stable suggests that, though the hybrid young are fertile, they are less so than the pure forms. The actual zones of hybridisation have changed over the years, with the Carrion Crow pushing north-westwards in the British Isles. The Hooded Crow is mainly resident, though the populations from the north-east of the range migrate to winter in western Europe. These movements are rather irregular and irruptive in character, probably reflecting local climatic conditions.

Migration: Autumn movements mainly October–November, further movements take place through the winter depending on climatic conditions. Return passage takes place in February–April.

Length:	47 cm
Wing length:	33 cm
Weight:	400–800 g
Voice:	Like Carrion Crow
Breeding period:	Beginning of April to May. 1 brood per year. Replacement clutch possible
Size of clutch:	5 (4–7) eggs
Colour of eggs:	Not distinguishable from eggs of Carrion Crow
Size of eggs:	41.5×29 mm
Incubation:	As for Carrion Crow
Fledging period:	As for Carrion Crow

ɪrkey, 22.5.1974 (Li)

Young birds in nest, Seewinkel, Austria, 1966 (Li)

Schleswig-Holstein, 10.4.1979 (Li)

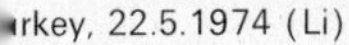

ɪft) Ad., Turkey, 1.6.1974 (Pf)

Raven (Corvus corax)

The Raven inhabits remote and undisturbed areas such as mountainous country, sea-cliffs and coastal islands, extensive areas of open woodland and other areas where man seldom disturbs them.
The diet consists of carrion of all types, eggs and young of other birds, large insects and larvae, small rodents, fish and edible refuse near human habitation. It feeds on the ground and is usually seen singly or in pairs except when food is plentiful, such as at a garbage dump or at large dead animals. It roosts socially in some areas.
It is a large, powerful species, the largest European passerine. It performs acrobatics when displaying, usually in twos or threes. It is a shy and wary species due to continued persecution by man, which has resulted in a reduction of range in some areas.
The nest is placed in a large tree, though exceptionally in a bush or small tree, on cliff ledges, on ruins or rocky crags. The nest is a bulky structure of sticks, thick plant stems, roots and other plant material, mixed with some earth, heather or green foliage and linen with wool and hair. Often bleached bones are incorporated into the nest. Some pairs have several different nests which are used over a period of years. The species is faithful to the nest and will build it up over the years to form a massive structure. Incubation is by the ♀ alone with both parents bringing food for the young.
The Raven is mainly resident and sedentary, though it may wander in winter or if food is in short supply. Sometimes flocks of Ravens are seen moving in the northern part of the range but no regular migration takes place.

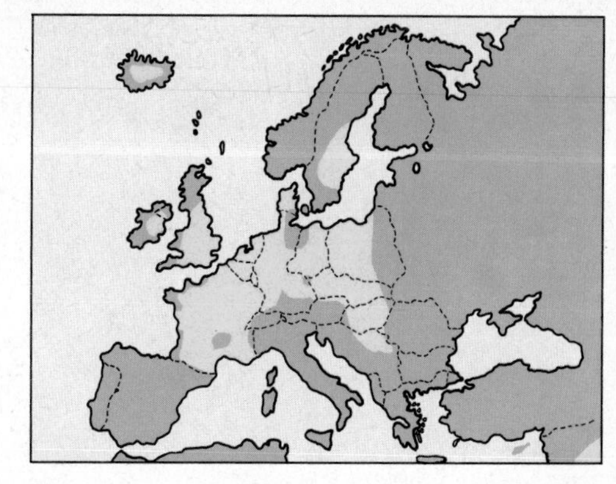

Length:	64 cm
Wing length:	44 cm
Weight:	*ca* 1250 g
Voice:	Hoarse, deep 'kruuk', 'kronk' or 'tok'
Breeding period:	February–April. 1 brood per year. Replacement clutch possible
Size of clutch:	4–6 (3–7) eggs
Colour of eggs:	Greenish to bluish ground, with dark brown or olive spots
Size of eggs:	50×33.5 mm
Incubation:	20–21 days, beg. before clutch complete
Fledging period:	Nidicolous; leaving nest at *ca* 5 weeks

ɫ. calling, Bavaria, 21.4.1973 (Li)

Nestlings, Bavaria, 1.4.1973 (Li)

Bavaria, 8.3.1973 (Li)

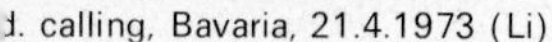

ft) Arrival at nest, Bavaria, 2.4.1973 (Li)

The Photographs

The abbreviations in brackets in the picture-captions indicate the names of the photographers who have supplied their often irreplaceable photographs for reproduction in *The Breeding Birds of Europe*. Addresses are in West Germany except where otherwise indicated.

(Aich.) Prof. Ambros Aichhorn, Salzburg/Austria
(Becker) Peter Becker, Diekholzen 2
(Bey.) Horst Beyer, Bielefeld 15
(Bosch) Johannes Bosch, Karlstadt
(Br.) Dr Peter Bracht, Reisach
(Ca.) Aldo Casata, Böblingen
(Chr.) Arthur Christiansen, Rödovre–Copenhagen/Denmark
(Di.) Jürgen Diedrich, Ronnenberg 1
(Diep.) Georg Dieplinger, Pietling
(Fe.) Walter Fendrich, Vienna, Austria
(O.v.F.) Prof. Otto von Frisch, Braunschweig
(Gé.) Benny Génsbøl, Naestved
(Gl.) Hans Glader, Rhede
(Gr.) Rudolf Großmann, Wilhelmshaven
(Haas) Dr Dieter Haas, Tübingen
(Hau.) Hannu Hautala, Kuusamo, Finland
(He.) Siegfried Hemerka, Klagenfurt/Austria
(Kühn) Manfred Kühn, Dachau
(Kü.) Karl Kühnel, Dietramszell
(Li.) Alfred Limbrunner, Dachau
(Mal.) Peter Malzbender, Wesel
(M.-D.) Johs. Meher-Deepen, Spiekeroog
(Par.) Helmut Paetsch, Wemding
(Pf.) Manfred Pforr, Moosburg
(Pl.) Alfons Plucinski, Goslar 1
(Qu.) Georg Quedens, Norddorf/Amrum island
(Rei.) Fred Reinwarth, Dachau
(Sch.) Otmar Scharbert, Bürgstadt
(Scher.) Dr Wolfgang Scherzinger, St Oswald
(Schu.) Werner Schubert, Sindelfingen
(Schw.) Karl Schwammberger, Oberstenfeld
(Sieb.) Rolf Siebrasse, Bielefeld 1
(Sier.) Manfred Siering, Grünwald
(Sw.) P. O. Swanberg, Falköping/Sweden
(Sy.) Gunther Synatzschke, Rotenburg/Wümme
(Ta.) Hugo Tannert, München 50
(Tö.) Konrad Tönges, Lahntal-Großfelden
(Wa.) Dr Klaus Warncke, Dachau
(Wo.) Konrad Wothe, München 81
(Wü.) Klaus Wüstenberg, Radolfzell 16
(Zeim.) Kurt Zeimentz, Ruhpolding
(Zm.) Jakob Zmölnig, Rotheenthurn/Austria

Index
of English Bird Names

Distribution Maps

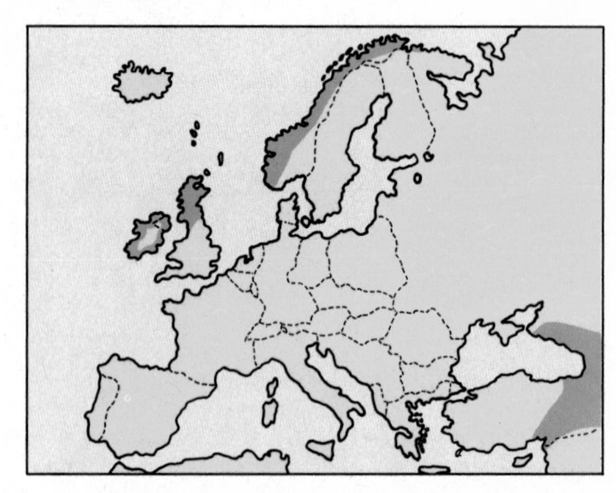

The green area shows in each case the breeding area of one species. This does not imply that that species is to be found breeding everywhere within the area indicated. A precondition is a suitable biotope (habitat). To simplify localisation, political boundaries have been indicated. Isolated cases confirmed outside the main breeding area could not be included owing to the size of the maps. In any case such cases are generally irregular and may die out rapidly. On some distribution maps two species have had to be included, in which case the breeding area of the second species has been indicated with a black dotted area.

Distribution maps: Manfred Pforr

Symbols and Abbreviations

The following abbreviations and symbols have been used in the text:

Ad. (adult) = sexually mature bird, adult plumage
Imm. (immature) = not in full adult plumage, sexually immature
Juv. (juvenile) = juvenile plumage
BP = Breeding plumage
NBP = Non-breeding plumage
♂ = male, ♀ = female

Index

of Scientific Bird Names

Bibliography

Handbooks Covering European birds

D. A. Bannerman and G. E. Lodge, *The Birds of the British Isles* (12 vols., London, 1953–63).

S. Cramp *et al.* (eds.), *Handbook of the Birds of Europe, the Middle East and North Africa: The Birds of the Western Palearctic* (Oxford University Press, Oxford, 1977, 1980).

G. P. Dementiev and N. A. Gladkov, *Birds of the Soviet Union* (6 vols., Jerusalem, 1951–4).

H. F. Witherby *et al., Handbook of British Birds* (5 vols., Witherby, London, 1940–3).

Identification

B. Bruun and A. Singer, *The Hamlyn Guide to Birds of Britain and Europe*, 2nd edn (Hamlyn, London, 1978).

H. Heinzel, R. Fitter and J. Parslow, *The Birds of Britain and Europe with North Africa and the Middle East*, 4th edn (Collins, London, 1976).

L. Jonsson, *Penguin Nature Guide Series* (4 vols., Penguin, London, 1978).

R. T. Peterson, G. Mountfort and P. A. D. Hollom, *A Field Guide to the Birds of Britain and Europe*, 3rd edn (Collins, London, 1976).

Migration

B. E. Bykhovskii (ed.), *Bird Migrations: Ecological and Physiological Factors* (1973).

R. Durman (ed.), *Bird Observatories in Britain and Ireland* (Poyser, 1976).

G. V. T. Matthews, *Bird Navigation*, 2nd edn (1968).

R. E. Moreau, *The Palaearctic–African Bird Migration Systems* (Academic Press, London, 1972).

G. Zink, *Der Zug Europaischer Singvogel: ein Atlas der Wiederfunde Beringter Vogel* (2 vols., Moggingen, 1973–5).

Biology and Behaviour

D. S. Farner and J. R. King, *Avian Biology* (Academic Press, New York/London, 1971–5).

D. Lack, *Population Studies of Birds* (Clarendon, Oxford, 1966).

R. K. Murton and N. J. Westwood, *Avian Breeding Cycles* (1977).

Distribution

J. L. Peters *et al., Checklist of the Birds of the World* (Harvard University Press, New York, 1931–64).

J. T. R. Sharrock, *The Atlas of Breeding Birds in Britain and Ireland* (B.T.O., Tring, 1976).

C. Vaurie, *The Birds of the Palaearctic Fauna* (2 vols., Witherby, London, 1959–65).

K. H. Voous, *Atlas of European Birds* (Nelson, London, 1960).

Monographs and Species Groups

L. Brown, *British Birds of Prey* (New Naturalist, Collins, London, 1976).

L. Brown and D. Amadon, *Eagles, Hawks and Falcons of the World* (2 vols., Country Life, London, 1968).

J. Delacour, *The Waterfowl of the World* (4 vols., Country Life, London, 1954–64).

D. Goodwin, *Pigeons and Doves of the World* (British Museum (Natural History), London, 1967).

D. Goodwin, *Crows of the World* (British Museum (Natural History), London, 1976)

W. G. Hale, *Waders* (New Naturalist, Collins, London, 1980).

I. Newton, *Finches* (New Naturalist, Collins, London, 1972).

C. Perrins, *British Tits* (New Naturalist, Collins, London, 1979).

E. Sims, *British Thrushes* (New Naturalist, Collins, London, 1978).

General

R. S. R. Fitter, *Collins Guide to Bird Watching* (Collins, London, 1963).

J. Gooders, *Where to Watch Birds in Europe* (André Deutsch, London, 1970).

J. Gooders, *Where to Watch Birds* (André Deutsch, London, 1967).

S. Smith, *How to Study Birds* (Collins, London, 1945).

A. L. Thomson, *A New Dictionary of Birds* (Nelson, London, 1964).

Journals

Ibis (British Ornithologists' Union).

Bird Study (British Trust for Ornithology).

Ringing and Migration (British Trust for Ornithology).

Irish Bird Report (Irish Ornithologist's Club).

Annual Report (The Wildfowl Trust).

Birds (The Royal Society for the Protection of Birds).

Scottish Birds (The Scottish Ornithologists Club).

Dutch Birding (Holland).

Recordings

P. J. Conder *et al., British Garden Birds* (Record Books, London).

J. Kirby, *Listen . . . the Birds* (European Phono Club, Amsterdam).

V. Lewis, *Bird Recognition: An Aural Index* (HMV, London).

M. E. W. North and E. Simms, *Witherby's Sound Guide to British Birds* (Witherby, London).

S. Palmer and J. A. Boswell, *A Field Guide to the Songs of Britain and Europe* (Sveriges Radio, Stockholm).

J. C. A. Roche, *A Sound Guide to the Birds of Europe* (3 vols., International Centre for Ornithological Sound Publications, Aubenas-les-Alpes).